Karl Heinz Johannsmeier

WOODEN BAZOOKA

Karl Heinz Johannsmeier

WOODEN BAZOOKA

From secret police prisons to success in Silicon Valley

Andromeda Publishing

For information address Andromeda Publishing,
611 Washington Street Suite 2103 San Francisco, CA 94111

Book and Dust Jacket design: Andromeda Publishing
Printing: Walsworth Publishing Company, Marceline, MO

Printed in the United States of America

ISBN 09675016-0-1

Close to Zero

I was fifty-three years old and had only seconds to live. I stood suspended above an abyss, supported only by my skis wedged against two vertical ice walls, while my eyes followed the huge block of snow that had just sagged away from under me, down, down into the dark crack below. Once the loose powdered snow had settled, my eyes could follow the greenish blue ice walls fifty meters down to where they disappeared in darkness. I had not heard the snow block hit bottom, so the crevasse had to be even deeper. All I could think was, "This is the end. You have not made a will, and you are leaving without saying goodbye to your family and friends."

When the surface had begun collapsing under me, I had whirled around on my skis and frantically tried to run back to level ground. But the block of snow beneath me had begun sinking like an elevator, and I had scraped with my skis to dig in along the walls it was leaving behind. Then the whole snow block—at least two meters thick, ten meters long, and over one meter wide—had fallen completely away, and now all was empty below me. I stood wedged in the gap it had left, with my legs spread rigidly apart, stemming my skis in the soft sides of the snowy upper walls, dead center in the crevasse about five feet below ground level. I was barely able to peer out over the surface and see Ingrid, my skiing partner. All I could do was keep myself perfectly balanced and still as a statue. And wait until the soft snow under my skis would give way and my long fall would begin.

My choice seemed to be to wait and let nature take its course, or to initiate the fall myself and have it end all the more quickly—controlled and clean. Maybe there might be a lot of soft new snow down there that could soften my fall. No; after falling more than fifty meters, I would be killed. Anyhow, I knew most of these cracks narrowed near the bottom into a wedge; after smashing against the walls as I fell, it wouldn't matter whether the bottom was soft. In just a few seconds I would find out whether life after death existed.

It was terribly cold and snowing like crazy. The wind kept blowing the low, misty, snow-loaded cloud along the freezing surface right into my face. Through the blustering snow I could barely make out the shadowy outline of the T-bar lift just twenty yards away. The increasingly severe weather had led us to ski away from this lift in the middle of the run a few minutes earlier. I had been afraid to ride on, because I figured it would be impossible for us to find our way down the mountain—we would get snowed under and freeze to death. Nobody else was still skiing and only this relatively primitive lift continued in operation, probably so that the ski patrol could sweep the mountain one last time for lost skiers.

All this had happened because I had wanted just one more run in the new powder snow the storm was dumping. I was skiing with Ingrid, a wonderful young German woman I had met the night before when she was modeling fur coats in the hotel where I was staying. We had talked after the fashion show and decided to spend this day skiing together. She was a great skier and turned out to be a great companion as well. In the sunny morning we had skied the Monte Rosa glacier.

After lunch we had come closer to the Matterhorn to ski, but the pleasant weather had retreated before a snowstorm. When we were only halfway up the *Trochenersteg* lift, the visibility dropped to almost zero. We saw a thirty-yard stretch between the lift and the *piste* (a slope for skiing) and decided to leave the lift and ski over. I had asked Ingrid to wait until I had skied across the somewhat uneven-looking area beyond the thin green-yellow rope to the *piste*. During the last few years I had skied only in California, where there were no glaciers, and had forgotten about their special hazards and the signs that warn of them, such as that rope. When I had skied less than ten yards beyond the rope, the surface around me began sinking. Confident and a good skier, Ingrid had disregarded my warning and followed not far behind me, but higher up, where the ice was apparently more solid. When I had felt the snow sinking, I had yelled, "Go back!" and she had turned and made it back beyond the line. There she now stood facing me, just ten meters away.

All I could say was, "This is where my life ends; give my final regards to my family in California." I knew I had to be quick with my words; but what came out sounded like a postcard, no match for the finality of the situation. But I found no other words. These would be my last ones, and all others, together with all the knowledge I had acquired in my life, would die with me. My mind was working frantically, while my body, stiff with fear, was locked into a fixed, immovable state. I looked down at my skis and noticed they were creeping very slowly downward and sideways. I did not dare look down too long because I felt dizzy. I tried to figure out how far I could sink before I reached the lower edge where the snow ended and the hard, slippery ice surface started. At that point I would fall for sure. But then, for some reason, I stopped sinking; now I tried to estimate how long I could keep myself balanced and how long my legs could support my body in this awkward split. I began to assess my position analytically, as an engineer. It helped that my ankles were fixed firmly into the ski boots, each forming together with my leg one big solid limb up to my hips. My two legs made me a stable pyramid, and if the snow against my skis did not give way much more, and if I did not slide, or bend forward or backward, or lose consciousness, and if the bindings didn't open but stayed fixed, I could last several more minutes. But what good would it do? There was absolutely nobody coming up the lift, and nobody skiing down the mountain from above. I thought of asking Ingrid to ski down the six-meter-wide stretch beside the lift to get help, but then I realized this would be futile.

The run was just too treacherous; she would not be able to make it down, and any help would come too late anyway. Strangely, I had not much to say. This was a hopeless situation, and I stood quiet with my mind now wandering, waiting for my end to come.

It seemed ironic that my life would end like this, here and now. I had just emerged from six years of uninterrupted, pressure-filled work, having finally achieved my dream of success. I had started and built up a high-tech company in California's Silicon Valley. I had just sold it, and, having achieved financial independence, I was setting out to do things I had always wanted to do—paint, write my memoirs, live a bit better. This vacation to Zermatt was supposed to be more glamorous than previous ones, somewhat like the lives of the rich and famous shown in *Town and Country* magazine.

But these dreams and hopes were now going to be snowed under. I could hardly see the big colorful diamonds on Ingrid's bright, Picasso-like ski outfit just ten meters away. A thick layer of snow now lay on me, and I could no longer see my own skis.

The seconds were ticking away into minutes, but I was still locked into a countdown to my death. Nobody was going to come up here in this weather. We were as alone as if we had been on a polar expedition. I thought that Ingrid would be the last human being I would see—she would not leave until I lost my balance and fell out of sight. I also thought of my father and wondered if I would meet him on the other side. He had died when he was just twenty-nine years old, and it seemed funny that in such an encounter I would be twice his age.

I had looked death in the eye before, but this was different. This was not a danger that was over before I had fully realized what might happen, nor was it the situation in which a person hampered and dulled by disease learns from his doctor that he has a certain window of time. With all my senses at their keenest, this encounter with finality was incredibly immediate, perfectly clear, and absolutely hopeless.

I began to count all the times when I could have died—the many diseases I contracted in childhood that could have killed me, the time I almost drowned, the time I fell through the ice into the river, the time a bomb knocked me unconscious during the war, my time as a guest of the secret police, my escape from East Germany, the airplane flight when fire trucks and ambulances waited on the runway, the amoebic dysentery I contracted in India. I concluded that I might be on my ninth life, but that I could not really apply what was said to be true for cats to myself.

Even so close to it, I could not actually visualize death. There was nothing to hold onto, nothing manifest that resembled anything humans had made up of him or it, no grim figure assembled of bones, and no special view to the other side. Here, at least, death was hidden, waiting in the dark below the blue-green ice walls, shadowy as the space into which I would fall any moment.

I decided, in my wait for the inevitable, that except for the actual fall, this would not be a bad way to die. Yes, it had been stupid to ski beyond the line, but otherwise this was a most unusual, satisfying, and special death, appropriate for an adventurer like me. I would just disappear: they would never recover me and I would never be buried. I would get deep-frozen and slide through thousands of years to the end of the glacier and into another time. Perhaps at the end of this long journey I might even be found well preserved, and possibly be brought back to life. Even today, new technologies existed that could revive hypothermia victims. What would the future bring? The only real loss would be the few remaining years of my present life, but that loss might be repaid by a trade-off, because in 3,000 years, technology would certainly have improved enough to make choosing the length of one's life possible. Why had nobody ever thought about that before, and drawn the obvious conclusion—kill oneself now, at a good comfortable age, and continue one's life almost indefinitely, later on? Best to do it young, because it would always be very difficult to make someone young again after his body had degenerated with age.

I had now endured several minutes in my locked position; because of the cold I did not feel any pain or cramps. I continued to think about life suspension. Before I founded my last high-tech company, I had wanted to do something easier. I had been tempted to start a low technology company based on the premise that preservation of one's DNA might give a person the opportunity to live again in the future.

But my life was over, and all these thoughts were ridiculous. Nothing mattered anymore, and I began again thinking about the impending end, the merit of the life I had lived, and whether I was going to heaven or hell. I had concluded years before that no one could know for sure where he was going, regardless of how good a life he had lived. I had always tried to be a good guy; nevertheless, there were episodes I had often wished to erase or be forgiven for. In the process of making myself, I had made some mistakes. Never had I gone to a priest for confession, though, or asked the authorities to punish me. I had just carried on and tried to improve. Over time

I had buried bad actions under good ones or had simply forgotten about them. Now, faced with immediate death, I could not tell whether the sum of my actions would send me to heaven or hell. When I looked again along the ice walls down into the dark hole below, I got the impression I would, like most people, just fall in between.

I had stood now for more than twenty minutes between these snowy walls. I had not sagged any deeper, and the wind had blown a drift of snow into the crevasse, partially covering it and me. I figured now that before I fell, I would probably freeze to death, or perhaps not fall at all, but just get frozen into the glacier right here.

At this moment a single skier appeared on the lift, a man from the Swiss ski patrol with a big red cross on his chest. While skiing out of the lift, he yelled over at me, "What are you doing over there beyond the line?"

I was searching for words to explain, but he had already fired the next question, "What is your name and in which hotel are you staying?"

That question sounded as if he were about to reserve a place in Zermatt's graveyard for me. I did not answer, but asked instead, "Is there any way you can help me?"

He replied, "No, I can't; nobody can get you out of there!"

His negative attitude kick-started my brain into searching for new ways out. The glacier rope. I knew it was too thin to support my weight in a fall, but it might be strong enough to keep me suspended where I was, or hold me against the ice wall as I climbed out.

I said, "I want you to loosen the glacier rope below, leave it fixed up there, and try to throw one end of it to me."

His answer stunned me and at the same time made my blood boil: "It is against the law to loosen the line."

But Ingrid was already on her way down to untie the knots. He tried to prevent her, but when she started to beat him with her ski poles, he let her do it. Ingrid then came up toward me and started throwing the line herself. Because I was carefully maintaining my precarious balance, I did not dare to move more than my hands and forearms to catch it; and each time the line came close, I felt vertigo and extreme fear of falling. None of it mattered anyway: from a distance of ten meters, Ingrid could never throw the line close enough.

Then I begged the ski patrol man to walk just a few feet to where Ingrid had been before and throw me the line from there. He did not want to do it at first, but finally he took the line and stepped a little closer. He threw it once, twice, missing both times. Then again, and again.

It took ten more attempts, but at last I caught the line. Now there was a chance. I let my poles fall into the crevasse and tied the glacier line around my chest. It was a very thin line, but if I could get to the snow wall beside me, the friction of my body against the wall would reduce the pull on the line by a factor of ten. With the line just keeping me from a fall, I might be able to climb out. I told the ski patrolman to take all slack out of the rope. I would lunge toward the snow wall on his side, and he should hold the rope tight, but not pull too hard.

I took a deep breath and leaped over the gap; my hands grabbed, dug in, clung to the snow wall. As he held me in place with the line, I inched my way up the crevasse, slowly pulled myself over the edge, and then crawled, my hands pulling my body, legs, and skis back to the lift. My knees were shaking so badly I could not stand up for some time.

I thanked the ski patrolman many times. I asked his name, too, but even though it was important, I forgot it right away. Strangely, I found myself feeling extremely distant from the two people who had rescued me and were now trying to treat my frostbite. The few words I managed to get out sounded disoriented and hollow. Having just returned from certain death, I was still in another world.

Eventually I recovered somewhat, and we started skiing down the narrow, treacherous stretch around the lift. We skied from lift post to lift post, stopping at each one. When we were almost down, Ingrid took a bad fall and could not lift her arm. We continued all the way down and found a man with a snow cat to drive her down to the main lift while I skied there. Later, when we got to the village, a doctor took one look and said to Ingrid, "Your upper arm is broken clear through." He put her arm into an orthopedic sling to pull the bone ends together, and let her go.

I decided to pack up right away and drive Ingrid to her home in Stuttgart. It was still snowing like mad, so we had a long, difficult trip. While I drove, I kept thinking about my accident and her broken arm, but I tried to distract her from her painful injury by asking questions about her life, and telling her about mine.

She told me she had not had much of an education and was suffering with a very jealous boyfriend who always found fault with her manners and kept correcting her speech. I told her this was somewhat typical of young men and resulted from insecurities. As a young man, I had been terribly jealous myself and had even made one of my girlfriends ask her previous acquaintances to return all her pictures. With another girlfriend who had not been able to learn much in school, I had tried constantly to improve her language skills and had given her German lessons twice a week, even though I had often flunked tests in spelling and grammar myself. I told Ingrid that during one's life one could learn many things one had missed in school.

She talked about how difficult it was for a woman to

get old; she did not really know what to do after her modeling career came to an end, but she might try to become a photographer. I advised her to sort out her talents carefully and then concentrate on just one, in order to gain some leverage—the world was always full of opportunities.

She asked whether I had always known what I could become; I told her that nobody knows in the beginning, and that things often change unexpectedly even in the middle or end of one's life, as they had for me. I had inherited the genes of my artistic father, but had been raised by a strict, conservative mother and a smart grandmother. I had seen how appealing it was to live a Bohemian life as a painter or a musician, sitting and talking lazily in cafes. But the genes of my hardworking grandmother had overpowered this inclination and I had ended up as an engineer, entrepreneur, and executive with a rather strong work ethic and with responsibilities I had never intended to shoulder. I had left forests and meadows behind, and had switched from using my artistic talents to my mathematical ones, even though this way of life was in many ways unsuitable for my soul. I had spent most of my life behind the concrete walls of factories, had used my creativity to invent machines I actually hated, had read too many technical books, had spent too much time with engineers whose struggle to keep up with fast-moving technology left them little time for art or literature—engineers who were driving our civilization into the future, giving everybody telephones, cars, TVs, and computers which were supposed to connect all of us more and more, but, in the process, were separating us ever further from each other.

By this time we had already driven down from the mountains and were sitting in our car riding through tunnels on a railroad train; Ingrid was sleeping. As I drove down the ramp off the train, she woke up and we continued our conversation. Because of her many questions, I did most of the talking: how I had lived as a child, how I made it through the war alive, how I had become a political prisoner, how I had escaped from East Germany, why I had emigrated to America, and how I had become successful.

I could not answer all her questions. There was not enough time, and Ingrid kept falling asleep. In order to stay awake, I kept talking to myself through the night until early in the morning when we ran out of gas on the *autobahn* close to Freiburg, Germany. I had forgotten about fuel. Fortunately, the very first motorist, obviously a religious man, stopped and drained some gas out of his tank into ours; and so we went on.

Ingrid kept asking me, "How could you have done all this? It seems unbelievable."

I answered, "Everything I have told you is true. Life is very long, and one can do many things and make many mistakes and still recover to accomplish almost anything, if one is genuinely interested, commits oneself, concentrates on one's goal, and persists. And if one stays flexible and dynamic, one can even live different periods of one's life in different styles, and raise kids, too."

When we finally arrived in Stuttgart, it was nine o'clock in the morning. After a few cups of coffee, Ingrid asked me to drive her to a doctor. When we arrived at his office, she surprised me by starting a negotiation with him: "I am a model and have to demonstrate furs tomorrow night in Munich. I cannot afford to lose my job. I want you to give me some injections so that I don't feel this stupid arm anymore." The doctor, as astounded as I, at first refused. I asked him to do as she said; it did not matter how much it cost, nobody would ever know. He ended up giving her the shots and pills she needed. The next day, Sunday, I drove her to Munich, where she modeled without anybody noticing her handicap. I thought, "With such concentration and courage, she can succeed." Soon after, I brought her to her hotel, thanked her again for saving my life, and said goodbye.

I could see she was deeply moved, and we hugged a long time. Collecting herself, she said, "I know I cannot allow myself to fall in love with you. I respect the fact that you're married and that you have a small child." She added, "I will probably never see you again. But if you ever write your memoirs, I would be very interested in reading them, and will keep you in my memory that way."

1

In 1933, when I was four years old, my father shot himself. He had just shot and killed his girlfriend, and had also shot her father and her brother.

I did not know my father well. My parents had divorced right after my birth, and my father did not visit us very often. My mother, my brother, and I lived in the very center of the old medieval section of Jena, a town in central Germany, while my father lived about an hour's walk away in a large garden cottage high up on the slope of the Hausberg Mountain.

That morning I was still half asleep in my bed, smelling the fresh bread that had just come out of the big oven in the bakery on the ground floor of our building, when I heard a loud knock on the door of our apartment and people talking excitedly to my mother. But then I drifted back into sleep as my mind mixed the familiar morning noises of the street—noises which had been much the same for hundreds of years—into my dreams. Then my mother came into my room and told me, "Get up and dress. We have to go up to your father's cabin." My brother Friedrich, who was three years older than I, helped me put on my clothes and shoes, and a little later we walked down the street, past the market square, over the Camsdorfer Bridge and up the Hausberg Mountain. I walked after my mother and brother, gradually falling farther and farther behind. What had happened had no meaning for me. I was too little to know much about life and death. I thought even dead persons were still alive somehow: they were people who had just moved somewhere else, like my father, who was always away, but could come back from time to time.

Finally, halfway up the mountain, my mother and brother stopped to wait for me. I could not walk any farther—my mother had to carry me. When I asked my mother what had happened, she said that some people had attacked my father and he had shot them, and had then shot himself. I did not understand very well, and just wished we were already up there so I could see him. The cabin stood in a wonderful garden next to the forest and there were many fruit trees nearby. I loved to play there. I could run around and hunt for fruit, and sometimes I would see wild animals.

My mother and brother kept talking. My mind tried to picture the scene—with victims lying on the ground or on the floor of the cabin. But when we arrived, there was nothing unusual to be seen except for buckets of blood (or what seemed to be blood) from the clean-up, and I went into the garden and began playing.

My father had not killed himself in his cabin. His girlfriend and her family had come there and had quarreled with him. After he shot them, he had gone farther up in the beechwood forest with his gun and his dog. He had sat there for some time in contemplation before he shot himself. When his body was found, his dog was guarding it.

Because I never saw my father's body, and because my father was never officially buried, he continued to live some sort of half-life in my mind. Only many years later did I learn that my father's body had ended up in the anatomy department of Jena's university, where medical students were curious about, and so anxious to dissect, a man who had ended his life and that of others this way. Nor did I know till much later that this murder-suicide became a celebrated case in Jena, the major story of its kind in that era.

During the time afterward, though, I heard nothing. My mother never talked about my father or the circumstances of his death. His picture was never displayed. To her, my father had become an unwelcome memory. To this day I do not know whether my father committed murder, was guilty of manslaughter, or simply acted in self-defense. I heard later that during the struggle, his girlfriend had jumped in front of her father, putting herself in the line of fire. I presume the authorities did little to investigate the circumstances. By committing suicide, my father had spared them the need.

I know that later in my life, I sometimes felt condemned to suffer punishments for my father's transgressions, and I would use the incidents surrounding my father's death and guilt as a yardstick to measure the severity of everything unpleasant I encountered.

Most of what I know about my father I learned from my maternal grandmother, who told me that she had liked my father very much, and wished he were still around: he had been an eccentric but good man who neither smoked nor drank, who lived on vegetables, fruits, and nuts, and who was loving and very creative. Born and raised near Frankfurt, he had learned the trade of chimney sweep. He decided early on, however, that this was not the profession he wanted. So one day he put on his sandals and a "Jesus" shirt—a long-sleeved white shirt that reached down to his ankles—and crisscrossed Germany on foot for almost three years, working here and there at odd jobs, never shaving his beard or cutting his hair. Eventually he found steady employment in Berlin as an artist-painter in the UFA film studio. I still have some old photos of him. One shows him in his chimney-sweep outfit. In another, he has very long hair, is wearing his Jesus shirt and sandals, and carries a walking staff that reached far above his head. A third shows him at the UFA Film studios in Berlin, painting a scene with a five-foot-long brush. After his odyssey my father returned to Jena, where he tried to make a living as a painter.

Later, after the divorce, he operated movie projectors at the Astoria and Capitol movie theaters. At that time, being a projector operator was a very prestigious, almost glamorous job, because movies were such a novel invention. In his new profession he must have had opportunities to speak with many of the good-looking girls who worked as ushers. One of them was the daughter of a university professor. My father fell in love with her, but the professor and his son disliked the relationship and threatened my father often. One such confrontation apparently led to the tragic events up in that mountain cabin.

While my father was painting, he had met my mother on a train to the town of Apolda, not far from Jena, where he was traveling to sell some handicrafts and paintings. He hired her to help him in his activities, they fell in love, and later married. I don't know much about their marriage. My mother was probably not flexible enough to accommodate my father's mystical attitudes and inclinations, and may have been too high-strung to be a good match for him. Even so, they had three children together: my brother Friedrich, nicknamed "Frieder," a girl who died shortly after birth, and me. I was born on February 28, 1929, the coldest day in Jena's history, when the ice on the Saale River was one meter thick.

When I was born, my parents lived in a small apartment on the fourth floor, right under the roof, in an old house on the most central street of Jena, Loebderstrasse. At the time even educated and skilled workers were often out of work. I don't know how poor my parents were then, but shortly after I was born my father moved out. My mother had no income and owed money for the furniture in the apartment. A charity provided her with a small amount of money each month. In my earliest memories we always had too little money for food, and almost never had enough money to buy coal to heat the apartment.

Only one clear scene of my father remains in my memory. I am sitting on his lap, and, together with my mother and brother, watching one of his experiments near the only source of heat in our small apartment, a little iron stove. Coal briquettes were expensive, so my father had made briquettes out of the coal dust that had fallen through the grill into the ashes. He mixed the coal dust with newspaper to form something resembling the pieces we bought from the coal merchant. I remember that they burned for a little while, but once the newspaper was consumed by the flames, only a little smoke came out.

With two children and no money, my mother, whose name was Helene, was always under severe strain. She worried about everything, frequently complained of being sick or having aches and pains, and often had hysterical reactions. She took masses of tablets for these ailments. She was quite strict with us. I remember being beaten often, with whatever came into her hand.

I also think she did not like me too much. I was the last child, born when her marriage was ending, and I was just a burden. My brother was much better looking. Next to him I felt like an ugly duckling. As I had to wear Frieder's hand-me-down clothes, I looked like a second-rate version of him. Too, he was generally in perfect health while I, unfortunately, seemed to be always sick. I was, or at least I felt like, the unfavored son.

The house in which we lived had been built and rebuilt and extended many times over several centuries. It had odd corridors and staircases that led to different sections, making it an unintended labyrinth. The apartments did not have hot water; ours had a single tap for cold. In the entire house there was no bathtub or shower. The only toilet facilities for all the residents were situated in a dismal part of the house, down two flights of stairs from our apartment, around three corners, then down a long, dark hallway and to the left. Five stalls were bunched there, each equipped with a bench with a lid, which covered a stinking hole plastered with new and old excrement. A pipe was supposed to carry the wastes down to a collection tank at street level.

Even the adults in the house didn't want to go there more than once a day—and never at night. We children found those dark sections of the house mysterious and incredibly frightening, even by day. The long stretch of hallway was especially terrifying. Halfway down it, you had to climb three steps, at which point another dark hall came in from the side. We would always wait at the beginning of this long hall and listen for a while to determine whether the coast was clear, no matter how urgently we had to go. Then, carrying a candle or a spirit lamp in one hand and clutching a big iron key in the other, we would run toward the steps; as we ran, the flame would be blown out, and we would shuffle forward quickly in the blackness until we bumped into the bottom step. Then we would jump up the stairs in a big leap, run to the door of one of the toilets, feel for the keyhole, insert the key, slip in, and turn the key from the inside to lock ourselves in. We were safe then; but while sitting there, we would carefully listen for any noise, anxious to escape back to the safety of the main staircase.

In the houses in our ancient neighborhood, all kinds of insects had found permanent nesting sites deep in the cracks. These insects in our house were kept particularly warm in their nests by the ever-active ovens of the bakery on the ground floor; they came out every night in droves to feed on us. When I went to bed, I would switch off the light bulb and wait a few minutes until the bedbugs had crawled up the wall next to my crib. Then I would switch on the light and smash the beasts with my thumb into thick red splotches.

My mother often bought some sort of coarse clay material with which we would try to seal the cracks in the wooden floor, from which we believed these insects emerged. My mother's brother, Uncle Helmut, would then paint over the clay with thick oil paint, but all these efforts were fruitless. Each morning, even following a day of caulking, we would have new bites on our bodies. The insect battles remained part of life for us.

In those days children were left alone frequently. There were no baby-sitters. No one worried about kidnappers or fires, even though wood stoves were lit every day for heating and cooking in every apartment, and any of the old, dry wooden houses could have burned down in a matter of minutes. I remember one night when I was just two years old and had been left alone. I was probably fighting the insects and could not sleep. I had to go to the bathroom, but could not climb out of my crib to use the chamber pot. So I defecated in my bed. Since I couldn't lie down or sleep, I pulled all the sea-grass stuffing out of the rotting mattress. In time, it got mixed with my stool. When my mother returned home and saw me, she screamed. I had covered myself entirely with this filthy mixture. Stunned and disgusted, she quickly took out one of the big pots in which she normally washed the laundry, filled it with ice-cold water, dumped me in and dunked me under, and washed me with the same gluey soap she used to wash the laundry and scrub the floors while I kept swallowing my own filthy brew.

My mind has a few blurred scenes left of my father, mother, and brother when I was about two-and-a-half years old. We were walking on the *Lommerweg,* a wonderful woodsy path along a small creek called the Leutra. I was sitting in a little four-wheeled wagon, being pulled. It was fall, and we collected oily *buch eggern* (beech-nuts), which were hard to open but tasted delicious.

In the Middle Ages that little creek had run everywhere through Jena. It was frequently dammed up, and the collected water was then released to wash all dirt and garbage from the town, as a flood from a powerful storm would do.

Another memory from this early age of two or three is of a Sunday when we were at a wading pool in Paradise Park, Jena's central park. The adults stripped off my clothes and made me go naked into the water, while everybody laughed and photographed me. I hated those pictures. They survived and were shown around in the family for years, and were a source of much embarrassment for me.

I loved to play in sand. One very hot mid-summer day when I was about three, workmen dumped a huge pile of sand two blocks away in a little square called *Nonnenplan*. The next day my mother could not find me. She looked for me all day and had almost given up when, late in the evening, she saw a lone child sitting on the huge sandpile with a thick woolen hat pulled over his ears. It was, of course, me, still working on tunnels and castles. When I had sneaked out of the apartment in the morning, for some strange reason I had put on a thick winter cap. I had played in the burning heat all day among other kids, and now was the only one remaining.

Our flat had one window overlooking the street. We spent much of our time looking out at what was happening below. There was always something worth watching—a horse refusing to do what its owner wanted, two dogs fighting each other or making love and then not being able to pull apart, a teamster making a delivery, or the police going after somebody.

The daily business of the town was fascinating to me. There was a man who lit and put out the gas street lights. People brought goods to the many merchants on our street in big baskets on their backs, in pushcarts, or in horse-drawn wagons. (Automobiles were still rare.) Women carrying shopping bags went from stand to stand and from store to store making their daily purchases. I loved to wake up in my bed in the morning and listen to the brooms of the streetsweepers scratching the pavement, the even pace of the footsteps of the farm women carrying their goods in big baskets to the market, and the trip-trap sound of horses pulling wagons down the street or trotting in front of a carriage. I often tried to visualize how the horses used their legs while walking or running, and figure out which hoof was making which sound, but I never could.

In the early 1930s, I could also watch street brawls from our window. The Communists and the Nazis fought each other frequently. I recognized the Nazis early on as the ones who wore the brown uniforms and often marched in formation, while the Communists were less organized and were dressed in civilian clothes—often cheap ones. My mother disapprovingly called them *"Proleten,"* condemning them for their bad manners, rough street culture, and radical ideas. One day my father, who was carrying a small table to us, was mistakenly attacked by a group of them. They smashed the table and used the legs as clubs to beat each other up. My father was able to save only one leg—I remember playing with it.

My family's attitudes about politics were communicated to us through comments they made as we watched street fights and later as we heard about new laws issued by the Nazi government. My mother wanted nothing to do with either the Nazis or the Communists. The Nazis were objectionable to her because they were too militaristic, but I think she felt that they attracted into their ranks more intelligent people than did the Communists, and that they were capable of organizing and administering things well, and of maintaining order.

One day, when my mother was glued to the window watching a fight between Nazis and opposing forces down on the street, I felt a terrible urge to go to the bathroom. To go all the way down to the toilets was out of the question because just the day before when I had gone there, the bakery apprentices had scared the living daylights out of me by coming out from the dark and mysterious side hall and pretending to be ghosts. No, I would not go there again. For emergencies we had a chamber pot, but neither my brother nor I could find it anywhere. I really had to go, and several times my brother asked my mother to help us find it, but she stayed glued to the window. Finally we found a similar-looking utensil: it happened to be my mother's frying pan. Frieder asked her, "Can we use this? Heinz has to go!" My mother, not turning around, yelled, "Yes, by God, go, and leave me alone for a moment." She later regretted her words very much, because I laid my business neatly into that frying pan while my brother held it by the big handle under my bottom. For this act we received a spanking that seemed to us much greater than that old frying pan warranted.

After that, my mother began giving me precise instructions: what I could and could not do, when I had to be home, how far away I might walk and play. At first I was directed to play in a little triangular square across the street from our apartment. It was surrounded by three-hundred-year-old houses. At its center was a small fountain—a little boy holding a fish that had water flowing out of its mouth. I loved this fountain. (It still exists in Jena, but has been moved to another location.) My mother could watch me from our window, and I always kept an eye on it in anticipation of an order from her. She just had to appear in the window and right away I would ask her if she wanted me to come up, or whether I could continue playing.

At this age, all store windows had magic for me. I could look again and again at the many items displayed, and study each one thoroughly. Just two houses away from us was a shop with a large display window full of all varieties of cutlery. I would lean my head against the glass and imagine how each item worked, how people would use these scissors and knives, where they would end up, and what life each one of them might have.

My favorite excursion was a trip to the show window of a nearby pet store. I was fascinated by the white mice, the turtles, and the green leaf frogs that lived there. If I got lucky, the old lady who owned the shop would call me in, permitting me to see the many birds and goldfish they had for sale, and perhaps even pet some of the rabbits and guinea pigs.

When a marching band strutted by or played in the market square, I got very excited. I loved the shining instruments and the sounds they made. I would run to get a little stick with a tassel on it and rush down onto the street. To the amusement of the people watching, and much to the annoyance of the marching band's conductor, I would strut in front of the band, lifting my baton up and down to the rhythm. I did the same thing when the military bands played on the market square in front of the *Hanfried*, a statue of *Kurfürst Johann Friedrich der Grossmütige* (Elector Johann Friedrich the Generous), who founded Jena's famous university in 1548. Much later that institution was given the name Friedrich Schiller University.

On weekends my mother often took me to Paradise Park, which extended along the Saale River and had many beautiful trees. Sometimes I dug there all day, building automobiles in the sand. These were just holes with a bench and a dashboard, but for me they became as real as the autos on the street, and I would drive them everywhere. The many beautiful birds and red squirrels in this park were so tame they ate out of my hand, and we always brought some bread crumbs for them and the gold carp. I also loved to feed the many black and white swans that made this place so beautiful and serene. I had no idea that later, during the war, this peaceful oasis would become the site of one of the worst experiences of my life.

2

While I was two, three, and four, I was often very sick, mainly as a result of a heart valve problem and malnutrition. I spent a lot of time in hospitals and caught many infectious diseases there—measles, scarlet fever, diphtheria, and finally tuberculosis. At times I could only walk a few steps at a time, and my mother had to carry me. During my bouts of illness, I always had the same recurring dream about dying, but no matter how much I tried to remember it the next day, I could never put together the details of that dream and how my death was actually occurring.

Once I was no longer spending so much time in the hospital, I began to live a life more like that of other kids. We were poor and lived on welfare, but my brother and I never felt poor or disadvantaged. My mother cooked very simple foods like milk rice, crepes, spinach soup, potato soup, flour soup, and potato pancakes. We ate meat only on rare occasions. Sometimes we had herring, which was then abundant and inexpensive. Frieder and I liked *matjes* (salted) herring and boiled potatoes best; but whatever my mother prepared, we had to eat, even the horrible spinach soup, which I hated so much it brought me close to vomiting. My brother and I were always hungry, however, and so even when we were served foods we did not like, we usually licked our plates perfectly clean.

Wednesday and Saturday were market days, when all the farm women came to town with their colorful fruits and vegetables. We played among the stands, and at the end of the day often went through the huge garbage pile to pick out fruits that had been discarded because they could not be sold. To us, they were perfectly fine; we just washed them in the big market fountain called *Bismarck Brunnen*, one of our favorite hangouts, cut the bad parts away, and ate them.

Another place special to us was the back door of *Café Glaser*, the most prestigious café in town, where we sometimes bought a big bag of *kuchenrander*, or cake crusts. These were all the crusts cut off the many different cakes baked and sold during the day. The café's clientele wouldn't eat them; only we kids knew that these were the best parts of the cakes. A full bag only cost three or five pfennigs, and sometimes we even got them for free.

Saturday was a special day for my brother and me because that was the day on which our grandmother, Berta, would often come to visit us. A widow whose husband had been killed in World War I at the Battle of Verdun, she lived on a small pension from the government. Often she took us to the market to buy us a fish sandwich, consisting of a roll with Bismarck herring (a marinated herring), onions, and pickles. It cost 15 pfennigs. When she was in an especially generous mood, she would buy us a *rostbratwurst*, costing the dizzying sum of 35 pfennigs. (A hundred pfennigs made up one Mark: one pfennig had a value of about one-fifth of a cent at that time.)

Sometimes Grandmother gave us each an additional five pfennigs. This was a lot of money for us—we could go shopping two or three times with this amount. We loved to go to a little toy store, which was run by an old woman. She knew we bought very little, but she was kind enough to tolerate us. I was most fascinated by the many tin soldiers and tin Indians. One unpainted figure cost one pfennig. I always bought those instead of the painted ones, which cost two pfennigs; I imagined them colored in the stories I created with them.

Another favorite place to explore was the Woolworth store, where we roamed around and looked at everything. The clerks, unsure of whether we belonged to a shopper, never threw us out. In winter this was the warmest place in town, and we sometimes needed to warm up. We never stole anything. Living so close by and just being among all those things whenever we wanted made us feel as if we had everything already.

There was even an inexpensive solution for our periodic cravings for sweets. The owner of the pharmacy sold an unusual item called *suessholz*, or sweet wood. This stone-hard, dried-out, sugary-tasting wood had to be chewed long and hard, but it always tasted good. It was, as I found out much later, nothing more than dried-out sugar cane. The man behind the counter had tons of it in his drawers and probably paid almost nothing for it.

For Frieder and me, the best day of the year was Halloween. All the kids in town were allowed to make rounds through the neighborhoods, reciting a little verse:

> Ich bin der kleine Andreas,
> Liebe Leute gebt mir was,
> gebt mir nicht zu wenig,
> ich bin der kleine König,
> Lasst mich nicht zu lange stehen,
> Ich will noch ein Häuschen weiter gehen.

("I am the little Andreas, Dear people, please give me something; don't give me too little, I am the little king, don't let me stay too long, I still have to go to another house.") People gave us all kinds of foods—cake, apples, candy, *lebkuchen* (a raised ginger cake with dried fruits in it), sardines, *sprotten* (sprats), herring, and even pickles and sauerkraut, all of which we either ate right on the street or put into our big sacks to take home. Most merchants loved or pitied us poor kids, and gave us a lot.

Every Christmas we had a small tree lit with real can--dles, sang Christmas songs together, and ate Christmas *stollen*, a German fruitcake with raisins and almonds, which my mother prepared herself and had baked in the bakery downstairs. (Most poor people had no ovens, and used those of a local bakery when they had anything to bake or roast.) The gifts we received were never lavish, but they brought us great pleasure. One of the most treasured gifts I ever received was a toy cart with horses. I made lumber out of kindling wood, loaded it on the wagon, and pulled the horses and their cart through the entire apartment again and again.

One Christmas someone gave us a bunch of puppets with a wooden theater for them to perform in. We played with the ensemble constantly, but each time my mother ran out of kindling wood, she would break off part of the theater. After the theater was gone, she began to snatch my wooden blocks, and then finally the puppets themselves. I still remember watching them burn. I cried especially about the fate of my favorite, good Kasper, who, I believed, was following the devil to hell.

That little theater also served us as a battlement when our mother tried to hit us with that painful bamboo stick everyone called the "yellow uncle." When we could no longer had the Kasper theater to hide behind, we took preemptive measures: we cut the stick in four places, and it collapsed the next time she tried to hit us.

Parents of that generation believed that the best way to be sure children learned to behave was to discipline them with a stick. Every kid we knew was acquainted with a yellow uncle. A beating hurt, but it was never uncontrolled or violent, and I believe no harm was done.

I actually had very few toys but it did not matter: just using my imagination and whatever playthings I did have, I could play happily all day by myself at home or outside in my favorite place, the little square in front of the house. Around it were a butcher shop, a pharmacy, a shoe store, and a clothing store. My mother did not shop in those stores because things were cheaper on the market square just one hundred yards away, and most of our clothes came from the *Winterhilfswerk* (Winter Relief Fund). When the Jewish shoe store, Bodina's, or the Jewish clothing store had shoes or clothing that hadn't sold, their owners would say to my mother, "Lenchen, we have something nice for you," and give her some items for almost nothing.

Lots of Jewish people lived among us. They spoke German as we did and lived lives that seemed as normal as ours. Nevertheless, I had the impression that my mother and grandmother were somewhat apprehensive when they bought from Jews. They didn't think of the Jews as being dishonest, but believed that they had superior business skills which would enable them to come out ahead in a deal. For example, I remember the last warm jacket we bought from the clothing store across the street. It looked nice, was very warm, and lasted for years. But it cost us so little that for a long time my mother kept wondering when some concealed defect would show up. I believe that my mother and grandmother couldn't do business with Jews without feeling some uneasy consciousness of difference, some sense there was a foreignness to them. So many Germans shared this predisposition that when the Nazis declared the Jews enemies of the Reich, few people protested: after all, people thought, the Jews weren't really Germans.

Of course, I was too young to be aware that a change had taken place in the government and that the Nazis had come to power, or to have any sense of what changes that development would bring. Soon, though, there was a noticeable shift in the nation's attitude that we, its youngest citizens, reflected: we kids started to play more with tanks and soldiers than puppets. We especially liked the tanks that could climb over everything and fire their guns in rapid succession. The popularity of war toys increased further as Germany regained its strength and military might. No one realized at this time, however, that the world had begun the countdown to World War II.

But things were changing rapidly, and some events should have put people on the edge of their seats. I remember faintly one scene. I was out in the little square in front of our house when some SA men[1] drove by in a truck whose cargo bed was full of men, some of whom were only wearing white shirts and black pants with suspenders, while others had on only their long white underwear. The truck stopped farther down the road, where it made a bend to reach the market square. There some other SA men pushed more half-dressed men toward the truck and loaded them in. People on the sidewalk stopped and looked on, but my mother pulled me off the street into the house, telling me the Nazis were picking up all Communists, people who were against the state, and some Jews. I wondered about this, but soon continued to play with my toys inside the house. Only later, when I had learned to read and write, and noticed that the shoe store I was still calling Bodina's had at some point become Salamander, and the clothing store where we bought my jacket now sold uniforms for the SA and Hitler youth, did I begin to realize what had happened on that day. Only then did the memory of that ugly scene crystallize into its meaning.[2]

The older I became, the more I suffered living under the shadow of my healthy and faster-maturing brother Frieder. Compared to him, I was never good enough; I was always trying to catch up, always the one to whom bad things happened. If a snowball was thrown, I was the one who got hit in the face. And so it seemed with everything. My suffering became even more acute when one day I pointed out a fellow who had beaten me up on the street. His name was Bierdimpel, but I mistakenly called him *Dierbimbel,* which meant nothing, but sounded ridiculous nevertheless. Everybody almost died laughing, and from then on I was often called by that name, or just "Bimbel," whenever I did something stupid. For me this name took on its own degrading meaning, and I was ridiculed and put down with it for years.

Admittedly, I did things that made everybody in the family wonder what future I might have and what I would do next. Yes, I walked down the wrong side of the stairs right after my Uncle Helmut, who had painted the left side of the stairway, told me to walk on the right-hand side. Not only did I ruin the paint job, I also left footprints down the staircases below.

Yes, when I was smaller I touched that man in front of the Woolworth store on his hunchback, asking my mother, "What does he have in there?" She pretended not to know me, but the hunchback only turned around, showing a friendly and intelligent-looking face, and offered to buy me some ice cream. Perhaps he saw that I simply loved to touch things, including every dog we met—even the most dangerous-looking ones. But my mother suffered on these occasions: she was a shy woman, and was also afraid of dogs.

Yes, I brought dead fish home when the city decided to fill in *die Lache,* an old branch of the Saale River which the tanners had used in the Middle Ages. Admittedly, the fish had been dead for a while, but to me they had merely lost the water they needed, and looked perfectly good to eat. My mother doubted they were safe to eat, and threw them away.

Yes, I may have appeared filthy after wading up to my chest in that ancient mud, but how else was I going to get the fish? After all, the big golden carps I had caught from Jena's famous Paradise Pond, with thread and bent pins from my mother's sewing kit and a few bread crumbs, had tasted delicious.

And didn't I always make little gifts for Christmas at Fraulein Klopfleisch's, she who always gathered a few of us kids together before the holidays and helped us make wonderful little objects out of bast, paper, cardboard, and wood? I really liked creating things from almost nothing, and giving them to my mother as gifts made me feel even better. Every little thing I gave her made her very happy.

When I was five, someone gave me a *Kaufmannsladen*, a little play store with an opening behind which I could pretend to be a merchant selling sugar cubes, beans, bonbons, and glass pearls. I often played with this storefront with a little girl my age. I met her when I was invited to play in a house not too far away from our apartment. Her name was Thea. When I met this incredibly beautiful girl, I was so taken that I could not think of anything but her. This was the first time I fell in love, and I prayed every night to be allowed to marry her after I grew up. Unfortunately, she came from a family much better off than ours and her parents eventually prevented her from playing with me. I never forgot her and kept her in my soul until her image was finally obscured by experiences I had and acquaintances I made in my teens. It is still astounding to me that at this early age such a strong attraction for the opposite sex can exist. The platonic nature of this experience also influenced me later during adolescence and on into adulthood.

Several other kids lived near our apartment. Of them, my closest companion was a boy I will call Bruno. He had been part of the dead fish caper, but his family ate his dead fish and didn't get sick. Bruno was growing up without supervision and did pretty much whatever he wanted. My mother allowed me to play with Bruno with great reluctance, particularly as she realized that whenever I played with him, something major, mostly bad, happened. With Bruno, I carried out important animal experiments, such as taking cats to the fourth floor and throwing them upside down into the cobblestone-covered backyard below. We loved seeing them turn around in the air, land beautifully on all fours, and walk away as if nothing had happened. When my mother found out, I received a big spanking for my part in this. The cats didn't seem to bear a grudge, and continued to play with us. Our respect for their talent made us become fond of them.

Bruno and I often visited Herr Graf, two floors below us. He was a jovial old man who enjoyed our company. He told us many interesting stories and frequently made Bruno and me wrestle each other. Bruno and I had similar builds and were about equal in strength and skill; we would wrestle for him forever. My mother eventually wondered why my clothes got so terribly dirty. After she found out, she forbade me to go there. One day Herr Graf asked us to wrestle for him again, and I told him about my mother's order. "No," I said. "My mother does not allow me to wrestle in your apartment because the dust lies two inches high there." One can only imagine what my poor, embarrassed mother heard from him afterward.

One afternoon I did not come home, and my mother ran desperately through the streets looking for me. I was playing with Bruno at the Bismarck fountain and had forgotten time. When I saw my mother approaching in the distance and looked at the *Rathaus* clock, I realized it was two hours after I was supposed to be home. I got scared and ran in a big circle around her, through the *Rathaus*, and homeward. I arrived before her. But I was so fearful of what punishment would come that I suffered an attack of diarrhea, losing everything through my pants while running up the four flights of stairs to our apartment. I hid in a little space right under the roof where wet laundry was usually hung, until my furious mother found me. I received as many blows as the staircase had steps, but she had to wash and wax the entire staircase.

I was often sent out to buy things my mother needed and sometimes made awful mistakes if she did not write down exactly what and how much she needed. One time she wanted two pounds of lard and sent me out for it. I kept repeating, "Two pounds of lard, two pounds of lard, two pounds of lard," while going down the stairs, and while standing in front of the pet store, where all my animal friends were eating their food behind glass and in cages. When I finally arrived at the butcher shop, after a thirty-minute excursion to all my favorite places around the neighborhood, I could not remember what my mother wanted, no matter how hard I tried. Instead of two pounds of lard, I brought home two pounds of yeast. My mother, who needed this lard badly, got very upset and called me names like *Dummkopf* (stupid-head) and *Nichtsnutz* (good-for-nothing). Fortunately for me, the butcher, who had helped me in guessing the wrong item, took back the yeast and gave me the lard.

Bruno and I roamed the whole *Altstadt*, the original old center of Jena, and, later, all the inner parks, graveyards, and surroundings up to the *Teufelslöcher,* or Devil's Caves, about thirty minutes walking distance from our house. We found these limestone caves fascinating and went deep into them one day, hoping to encounter the devil for whom they had been named. We played there late into the night, and got whacked for coming home late and very dirty.

I was always afraid to wear new or clean clothes. I knew I could not do anything in them or have any fun. New clothes were too precious, rare, expensive, and hard to wash. Everything at this time was washed by hand in mostly cold water, and I was fearful of my mother's predictable reactions to any spot, dirt, or damage. I felt more comfortable in rags.

Bruno and I tried once to play with Annekate and Siegfried, the other two kids who lived in my building. They were brother and sister and had an extremely strict father. Bruno and I had always considered them rather stupid, perhaps even mentally handicapped, and had little contact with them. Both of them wore ridiculous-looking wire-rimmed eyeglasses with thick lenses and spoke with bird-like chirping voices that sounded like the voices I had used for my puppets. For some reason, while playing in their apartment we got the idea of digging out the brittle caulk that sealed and held the window panes in place. Our intention was to replace the caulk with fresh putty. But when Herr Harold, their strict father, came home and saw that all his window panes were about to fall out onto the street, he became terribly angry. Without asking any questions, he took out the yellow uncle and started to hit Siegfried and Annekate so violently that I peed in my pants as Bruno and I made our escape.

Annekate and Siegfried attended a special school that helped them tremendously. When I encountered them years later, they had, to my surprise, grown up to be perfectly normal, healthy people. Bruno did not fare as well. When I was eighteen, I encountered him one night at a dance, and he confided to me that he had already had a brush with the law and might continue down that road.

Movie theaters, or kinos as we called them, played a major part in my childhood. After my father died, my Uncle Helmut managed to get his job as movie projector operator. He behaved as if he was the owner of the place, and allowed us kids to slip through the back door. We usually had to sit in the last row of the second balcony, but on special occasions sat in the very first row next to the screen. This made us feel very privileged. Of course, these were considered by most people to be the worst seats in the theater, but for us they were special.

Movies had entertainment value for me, but they were also a very important source of information: they were windows into another world, a cleaned-up world in which the people were always well-dressed, ate in fine restaurants, danced, drank good things, fell in love, and had nice kids. Going to so many movies that depicted the good life made it possible for me to create in my mind another world for myself which became as real as the one I lived in, but which paradoxically seemed unreachable.

I remember loving and finding hilarious the comedies of Laurel and Hardy. One day in Uncle Helmut's projection room I received confirmation that the world of movies was real somewhere when Laurel and Hardy came in, in the flesh, and I was introduced to them and shook their hands.

Like the Laurel and Hardy films, the majority of the movies that I was allowed to watch were quite harmless: people kissed, but only on the cheek. If there was any real kissing involved, my uncle brought me up into the projection room, and I watched the movie from a little window beside the projector. My uncle kept track of the movie and edited it for me by closing the shutter. I remember some of those movies even today, as well as the newsreels that preceded the features and informed us about current events; most often these spoke proudly of how Germany was rising from the ashes of World War I, and showed how strong a military power it was becoming.

I enjoyed my early childhood and the experiences I had in my home town as it was at that time, and am grateful that my mother let me roam around freely, in spite of the dangers there were to encounter. I learned to think quickly, improvise, and cope with almost any situation. I survived very well.

The worst thing that ever happened was that a bicyclist ran over me in front of the Rathaus. His handlebar hit me very hard on the chin, and I remember seeing stars in front of my eyes before I fell down and passed out. Then I came to and cried, attracting a crowd, until finally one well-dressed lady led me to the Spanish Orangery, a fine delicatessen and fruit store, where she offered me a choice between a piece of watermelon and some bananas. I was five-and-a-half or six years old, and had never eaten a piece of watermelon in my life, but I had often stood in front of that store window and admired those exotic melons with their rich red color. I didn't know what they were, how they tasted, or how one ate them. I could not imagine my mother ever shopping at the Spanish Orangery, and the watermelon looked expensive and unattainable to me. I chose the bananas because I thought I could get more of them. I had to wait for my first taste of watermelon until I was thirty-one and had come to the United States.

As far as I know, my mother did not have any social life during those years when I was young. Until I was almost seven, there had been only one man in her life, a mysterious-looking Hungarian doctor who must have entered our lives when I was about five. He was mysterious because he had very dark hair and a beard around his mouth, and as far as Frieder and I were concerned, he was from an unimaginably faraway country. As I remember, though, he spoke perfectly correct, accent-free German. I don't know where my mother met him, but he visited us occasionally and watched me over the arm of the couch as I played with my horse and wagon, puppets, and wooden blocks. He would read us stories, fables, and fairy tales. I loved to listen to him. He had a wonderfully deep voice, which was more hypnotic than that of anybody I knew. Its sound catapulted me into a dream-like, fairy-tale world, filled with interesting characters, sadness, tragedy, fortune, luck, and happy endings. My favorite stories were "Sterntaler" ("The Shower of Gold"), "Zwerg Nase" ("The Nose"), "Rumpelstilzchen" ("Rumpelstiltskin"), "Rotkäppchen" ("Little Red Riding Hood"), "Schneewittchen" ("Snow White"), and "Das hässliche junge Entlein" ("The Ugly Duckling")—for whom I felt, because of my situation, a close affinity.

The Doctor, as we called him, had one very annoying habit. After he had read a story to me, he would always ask questions about it, to make absolutely certain I had understood every detail. I remember particularly an interrogation about "Sternthaler" which was difficult for me. Nevertheless, this man was obviously a very kind person. He was very intelligent, well situated with a general practice in a better section of town, and was very interested in my mother. He asked her to marry him, and my mother considered the proposition for a long time. Unfortunately, she did not have the nerve or imagination to marry someone who was both a foreigner and of a class higher than ours; ultimately she turned him down.

This was the worst mistake she ever made, not just for herself, but especially for my brother and me. It left her no choice but to take a job as a low-level employee at Carl Zeiss, a big optical company in Jena. She ended up working there for more than twenty-five years. Her decision deprived Frieder and me of any opportunity for better schooling, including the chance to go to high school, as well as of the perspective, influence, and wisdom such a man would have brought with him. A short time later my mother met another man who turned out to be just the opposite of the Hungarian doctor, a man from the lower class, unpolished and rough, who drank and often beat her up. This man she married.

1 The SA (Sturmabteilung, or storm troops) was the Nazi party's militia, with brown shirts as its uniform.

2 When the Nazis came to power in 1933, less than 1 percent of the German population was Jewish. The Nazis quickly began "an immensely popular program of discriminating against Germany's Jewish citizens." The government kept increasing the severity of these measures, with two objectives: to drive Jews out of Germany and to keep their property. By the time Germany attacked Poland in September, 1939, about half of Germany's Jews had left, and most of the property of German Jews had been seized. Although Hitler announced to the Reichstag in a speech January 30, 1939, that in any new war the Jews of Europe would be exterminated, the policy of mass murder had not yet begun. (Gerhard L. Weinberg, A World at Arms: A Global History of World War II (Cambridge, 1994), 97).

3

My life in the lively inner town I had loved so much ended when I was six-and-a-half years old. My mother consigned me to an orphanage not too far from Jena. This was a horrible shock, by far the worst I had ever received.

My mother had threatened several times to send me there because I was such a difficult and strong-willed child, but I never believed she would do it. When she finally landed a job, though, she concluded that she could no longer take proper care of us, and made the tough decision to leave my brother with our grandmother and place me in the orphanage. Also contributing to this decision was a new attachment: my mother had found a man she wanted to live with and was trying to make room for him.

The orphanage was located in a villa in the countryside on the outskirts of Jena, close to a little village called Ziegenhain. A path and a single unpaved country road connected Jena with the orphanage and the village, which was surrounded by gardens, fields, and forests. Up until 1935, university students came there to duel for their honor, sometimes with deadly results, or to enjoy themselves talking, drinking and playing games. Hitler's government changed all that, curtailing the free life styles of the students as well as all freedom of thought for academics. Honor was redefined and a "clean" life imposed on Germany's youth, with no drinking and no smoking allowed, and no women to be thought of but good German mothers.

This orphanage was probably a fine institution with a devoted staff, but I hated it from day one, and my mind made it into a penitentiary for children where the needy inmates were forced to do all kinds of dirty work. The shock of having been left there by my mother made me an even more difficult character than I otherwise would have been. I was not a child who would accept the transfer of my mother's authority to the orphanage staff. I could not fit myself in and never got along with the staff, who treated every one of the fifty children there equally, and tried to force me into a mold to make me behave like all the others.

It didn't work. I lasted there only a few weeks. I don't remember much about that place, the other kids, or the staff, but I remember very clearly the day I was ordered to clean and shine thirty pairs of shoes. I looked at the long row of dirty shoes on the shelf, and that sight triggered my break with the system that had been imposed on me. I decided not to take it any longer and to escape into the nearby forest to start a life as a hermit.

I did not know that forest very well, and I was afraid of getting lost and of being attacked by wild animals. I had not forgotten the stories of Hansel and Gretel getting lost and of the wolf eating Red Riding Hood's grandmother. Having made good my escape, I was scared enough to build a little protective shelter of branches and twigs, in which I slept with the same feeling of safety a child feels after pulling his blanket over his head. I slept deeply until the sun woke me up. Sleeping outside was no hardship on a warm and beautiful summer night, and the many wonderful things I found to eat in nearby gardens made up for any discomforts and frights.

I never thought of returning to my mother: I thought she had dumped me there for good and did not care about me anymore. I had no idea that my whole family and the staff from the orphanage would be desperately looking for me, no sense that my mother would be crazy with worry. Yet as far as I know, the police were never informed of my disappearance. Parents loved their children, but in those days parents looked and waited for a lost child as one does today for a missing dog.

The director of the orphanage was outraged by my recalcitrance. I remember the exact words he used in speaking to my mother: *"Das ist ein schwer erziehbares Kind."* ("This is a child who is very hard to ..."—the word *"erziebahres"* doesn't translate into any one English word, but combines the ideas "educate and raise," "handle and control," and "make behave.")

I don't remember whether I spent one or two nights in the woods. But when it started to rain, I became uncomfortable and lonely, and went back into Jena. Some acquaintances of my mother's recognized me as I stood in front of Woolworth's; somewhere I had picked up a little broom which I carried over my shoulder.

Running away from the orphanage had been the right move for me, as it turned out. The orphanage did not want to take me back, and my grandmother took me in. I was almost seven; I ended up living with her for the next eight years. My brother Frieder got the worst of it—he had to take my place at the orphanage. But he stayed only six months; then my mother, who had moved with my future stepfather to another part of town, took him back.

From then on, I saw my brother only occasionally. This separation allowed me to gain self-confidence and to develop more freely. I saw my mother frequently, and she remained the main authority in my life, while my grandmother acted more as a caretaker. I got along very well with my grandmother, and she took good care of me; but living with her was a big change for me, and must have been a big change for her as well. I had a character totally different from that of any of the children she had raised before: I was strong-willed and demanding, and I needed much more freedom. She learned quickly that my actions were unpredictable—as in the case of my second escape from Ziegenhain.

Shortly after I ran away from the orphanage, my Uncle Helmut and my Aunt Dora moved to Ziegenhain. They were very happy newlyweds. A few months later they invited me to spend a few days with them. I was very happy to come. I felt close to my uncle because he had allowed me to see all the movies, and my new Aunt Dora was so beautiful I could never take my eyes off her. My uncle also raised rabbits (chinchillas), and I loved to play with them and give them food. I was very gentle with those soft creatures, and can still feel the sensation of holding them in my arms. But my uncle and aunt made one critical mistake that would test me anew. They told me one day we would have big hot dogs (a novelty at that time) for lunch. My mouth had already started watering when they added, "But unfortunately we could obtain only two." That meant there was none for me, and I would have to watch them eat theirs.

It was eleven a.m. I said nothing, went out to my rabbits, and, while pressing one of them to me, tried to beat down my pride. But thirty minutes later I was already near the place in the woods where I had spent my nights when I escaped from the orphanage. By then my uncle and aunt had missed me, and soon the whole family was searching for me again. Later that night I walked down to the bus station and took the bus back to my surprised grandmother.

My maternal grandmother was commonly called *"Mutter"* (Mother) by everybody in my family. She had been born and raised in Kaltennordheim, Roehn, a small village one hundred kilometers from Jena. Her school grades were the best I have ever seen—all A's in eight straight years of school. When she was still a young girl, she left her village for the town of Eisenach, where she worked as a maid in a fine upperclass household. She met a young glass blower from Sonderhausen, married him, and moved to Jena, where he found a job at Schott. They had three children: my uncles, Alfred and Helmut, and my mother. In 1917 my grandfather was drafted into the army; shortly thereafter, he was torn to pieces by shrapnel at Verdun.

My grandmother lived on a very small amount of money. She knew, like no one else, how to save and stretch money and how to help people in need. She knew how to do a great many things, and could work harder than anybody else. She knew how to treat most illnesses and nurse you back to health. Completely unselfish, she was the Mother Theresa of our entire extended family, and she exerted, almost by osmosis, a great influence on me.

We lived very economically in her little apartment on the Magdelstieg. We would burn only a single forty-watt light bulb. Every evening we sat for an hour in the dark because she wanted to save even some of the electricity this dim light bulb consumed. She said that one could talk well without lights and that it was good for the mind to sit awhile in the dark every day. We called this the *Daemmerstunde,* or twilight hour. One of her talents sometimes came out in these conversations: she could imitate people better than a trained actress. In the dark she mimicked voices; if the light was on, she would also imitate the person's walk or characteristic odd behavior of familiar individuals.

Her small apartment under the roof had walls that slanted down quite far into every room but the living room. We called the living room *die gute Stube* (the good room), and rarely used it. (All but the poorest families had a *gute Stube,* used only on great occasions. The best furniture sat there in solitude most of the year.) We lived mainly in *Mutter's* kitchen. Cooking and eating took place there, and it was the only room kept reasonably warm. The living room had a *kachelofen* (a tall, tiled stove), but that was lit only on two or three days in a year. Even at those rare times, once the oven had been lit, she would put single coals on at the rate of one an hour—just enough to maintain the fire. One could only get warm by leaning against that stove. In winter the apartment was so cold and humid that in the living room and the tiny bedroom, where my bed was right next to my grandmother's, the walls grew thick ice crusts. Every winter the ice peeled off more of the wallpaper. In the evening and during the night no fuel was put on the fire. We slept under huge coverlets filled with goose down and never froze, but getting into such a cold bed was not pleasant. We often heated bricks or made hot water bottles and slipped them into the beds half an hour before retiring. Then in the morning, when the whole place had turned into an ice cave, Grandmother lit a new fire in the kitchen stove with newspaper and wood.

Mutter's apartment had a flush toilet. The roof slanted down so far in that room that only small people could sit upright, but I was grateful not to have to go to the bathroom with the deep, stinking hole anymore.

Like my mother's old apartment, this one had only one cold water tap. It was in the kitchen. We washed our hands often, our faces every morning, and our bodies once every week, all in a small basin atop a three-legged stand. By today's standards this may sound appalling, but I never smelled anybody's body odor, and I felt just as clean then as I feel today taking two showers a day. Grandmother kept her place immaculately clean. Every day except Sunday she would clean and dust, and even wax the staircase leading to our apartment now and then.

I loved animals, and *Mutter* allowed me to keep and raise every imaginable variety in her little toilet chamber. Because of them the room sometimes stank badly, which must have been an offense to her cleanliness; she often complained about the smell. Worse, sometimes one would escape, run through the apartment, and hide under the furniture. One time a green leaf frog escaped. We could not find it again. A few days later, when my grandmother had finished her daily routine of waxing and polishing the staircase, she saw a fresh leaf on one of the stairs and yelled out loud, "I just polished the stairs a minute ago, and he has already dropped a leaf." When she bent down to pick up the leaf, it made a gigantic leap onto her shocked face.

I loved to watch my grandmother while she prepared food. She cooked and baked frequently. I learned a great deal about the processes of cooking and baking from her. I would often sit right next to her while she was working dough, and sometimes, at the moment she took her eyes away, stuff some of my marbles into it. She would scold me then, but actually she took this quite well, even though it was a chore to find and pick them out again.

The women around Jena are famous for baking the best cakes in all of Germany. On special occasions and holidays, my grandmother would prepare all kinds of gigantic, wonderful-tasting cakes. When one was ready to be baked, I would carry it on a huge, round plate on top of my head to the baker's big oven. When the streets were icy, this was quite a balancing act, especially if no ashes or sand had been spread. The cakes always tasted fantastic.

I had often seen my grandmother pour rum into her cake dough. Once, after I had seen her do this, I sneaked out, opened the pantry where she kept the bottle of rum, and drank some the way one would drink water. It burned like hell as it went down my throat, but otherwise had no immediate effect on me. Then I went off to the barber to get a haircut. While I was sitting in the chair, after the barber had cut one side of my hair with his electric clipper and was starting the other side, I began to see everything double and the room started spinning. I had no idea what was happening. I only knew I had to get out of there immediately and somehow make it home. Fortunately, I did, and my grandmother, believing I was terribly ill, put me to bed right away. My mother happened to drop by that day, and, after my grandmother told her how sick I was, she came to my bed. She bent over and was about to say something nice when she caught a sniff of my breath. She yelled, "This guy is drunk!" to my startled grandmother.

Every four weeks or so my grandmother washed, dried, and ironed all the dirty laundry that had accumulated during the month. Often I helped her with this big job. We used a laundry downstairs that served everyone in our apartment house. She allowed me to light the fire in the big wash-kettle stove where everything washable was boiled. After everything was washed, we had to wring out the water from each piece of laundry—bed sheets, towels, socks, and garments—with a hand-cranked wringer. Then my grandmother hung the clothes up on the drying lawn close by, and laid some of the white pieces out in the sun for bleaching. Every family in our part of the city went through this procedure known as "the big wash." The drying lawn was often completely covered with laundry from all the people who lived in the surrounding apartment houses; only rarely was it clear enough that we could play there.

Unlike laundry, shoes required daily attention. Every day we cleaned and waxed them. In Germany's weather they experienced a lot of wear and tear. Some of the streets near us had no pavement and turned into mud or dust depending on the weather. As a result, shoes got very dirty: as no one could afford to buy new ones, we made our shoes last for years by caring for them daily and having them repaired again and again. Later, after the war had started, it became very difficult to obtain shoes at any price.

Shortly after my grandmother took me in, I made my very first trip by railroad. We traveled to Kaltennordheim to visit my great-grandmother. This trip opened up a whole new world for a city kid like me. After the train ride, we stayed in a village where geese, goats, and sheep were still taken out to pasture every morning, often by kids my age. I ate lots of honey because a neighbor had beehives. I even rode on my great-grandmother's enormous pig. I also remember being attacked by a billy goat twice my size and having to fight the ganders or run away from them. My great-grandmother, who was at that time ninety years old, did all the work in the house and the field. She cared for everybody, including the animals, helped only by a semi-retarded man who was a remote relative left in her care.

After our return from the countryside, I slowly began exploring my new neighborhood. It was entirely different than the *Altstadt* where I had lived and roamed before. Uphill, just about three hundred yards away, one could find wonderful gardens in which people grew their own vegetables and fruits. Beyond these were beautiful meadows and fields full of wheat, barley, corn, and potatoes, and then endless forests. There were mice, hamsters, rabbits, and some deer to see, and wild strawberries, boysenberries, and raspberries growing, with what seemed to be wild apple, pear, and plum trees. When I saw all of this, I was very, very happy. I felt the way a captive lizard must feel being moved to a bigger terrarium, or regaining his freedom completely.

If I went down the hill instead of up, I found a different scene. The city had built a housing project there for poor families with ten or more children. The first time I wanted to walk by this building, a kid no bigger than me, but much tougher-looking, walked up to me and asked my name. Before I had a chance to answer, he punched me in the face and gave me a black eye. I had, unwittingly, made the mistake of walking down the wrong side of the street, in "their" territory. I never did that again; this little episode woke me up quite a bit.

It was not long before I had met and joined with the boys from the area nearby. We played all kinds of games in a little square below our house. When our ball bounced over the play area's little spiked fence, we had to climb over to get it. One time when my brother visited me and climbed over the fence, he slipped and ran one of those spikes clear through his hand. After that, we went around the long way.

During this period I began to learn more about my city and to notice more of what others were doing there. Sometime in 1935, the year I went to live with my grandmother, I remember walking with my mother to the market square to attend a Nazi demonstration. This scene sticks in my mind because it was the first and last demonstration I witnessed until I joined the *Jungvolk* four years later. Somebody gave a long speech, and then everybody sang the national anthem and the *"Horst Wessel Lied,"* the song of a Nazi hero. I can still hear my mother saying, "We are now supposed to lift our arm, *das ist jetzt der Deutsche Gruss* (this is now the German way of greeting)."

I did not understand what this was about and was impatient at waiting so long for nothing. It looked strange to me that people lifted their arm that way. Later, when I joined the *Jungvolk,* we too always sang the *"Horst Wessel*

Lied" right after the national anthem. I could never quite understand why Horst Wessel, a man who had been shot almost accidentally during a Nazi demonstration in Munich, could have been so important that he deserved to have a second national anthem about him attached to the real one. But the song was a masterpiece of Nazi propaganda and mass psychology. Even now, more than fifty years later, the jingle sometimes rings in my mind.

My impatience with these Nazi mass events may have come from a child's inability to comprehend, but it fit with the prevailing attitude and spirit of Jena. The city's traditions and the kinds of backgrounds its people had made Jena unreceptive to the Nazis' stiff, brutal, militaristic ideology and organization. When Hitler had driven through Jena, the townspeople had thrown stones at him. He never returned.

In the Middle Ages Jena had grown and expanded from its beginnings as a small, romantic village on the Saale River, surrounded by limestone hills with many vineyards and forests. The town was proud that for almost four centuries famous poets, musicians, philosophers, naturalists, and scientists had studied or taught at the university and lived within Jena's walls. Many of their names are still inscribed on homes, museums, and institutes.

Many academically trained people lived in Jena, working at the university, the many institutes, and the famous clinics. Until the early 1930s young people came to Jena to study from all over Germany, and from many foreign countries as well. The students had their own very lively and active subculture. The late afternoon scene in the inner town, with many students in their colorful caps lounging in chairs in the market square, debating endlessly over their beers, remains engraved in my mind as the constant backdrop to my outdoor play as a child.

In addition to its many academic institutions, Jena also had the biggest, most renowned glass and optical manufacturers in the world. Carl Zeiss Jena (the optical company) and Schott (the glass works) employed a large number of workers, many of whom were highly skilled craftsmen, often with technical training. They were not just ordinary workers, but worker capitalists: Carl Zeiss and Otto Schott had bequeathed their companies to trusts owned by the employees. These worker-owned companies were virtually unique in the world at that time. They embodied Jena's tradition of independent thought and action. In this liberal, educated community, thoughts and aspirations I would never have picked up in my own family soaked into me.

After Hitler's takeover in 1933, the atmosphere in Jena changed rapidly. Foreign students disappeared. German students were forbidden to belong to societies, and could no longer participate freely in the activities Jena's intellectual community had engaged in for hundreds of years.

Sometimes when I went grocery shopping with my grandmother, I heard grownups exchanging remarks that reflected the new order. "Did you hear that they took so and so away, and this man came back, and they released that man?" When they talked about the line the state had drawn at one-half Jew, what happened to quarter-Jews, half-Jews, and full Jews, it sounded to me, a seven-year-old, as if they were still talking about groceries[1]:

"Good morning, Frau Petri. What would you like today?"

"One pound of flour, half a pound of sugar, a quarter pound of raisins, one-eighth pound of coffee, and one quart of milk."

By this time, anyone who wanted any civic job had to have an "*Annennachweis,*" an official document stating that one had no Jewish ancestors and listing the names, birth dates, and birthplaces of one's ancestors for three generations back.

Nobody I knew was overtly against the Jews, and many, like my mother, expressed concern about the fate of Jews who were friends. But as long as nothing happened to them or theirs, most people were indifferent—rather like people who see a crime on the streets of a big city and avoid getting involved.

I was only a child then, but when I think back I believe that, like most people, I remained largely ignorant throughout the war of what was happening to the Jews. We had been told they were enemies of the state and they had disappeared from public life. If they were suffering, so were we, and many of "us" had been killed. With these and other thoughts we blocked our minds and kept ourselves from suffering from true feelings and guilt.

When the war ended and I saw the sickening and horrifying pictures from Buchenwald of prisoners in striped clothing starved to the very fringe of death and trucks and graves full of bodies, I had a hard time believing Germans had done all that. I wanted to hold on to the self-respect that being a German had always given me. I barricaded my mind from being blemished by concluding that our former enemies were making up stories to degrade us. I adopted the belief that the starvation had resulted from the breakdown of food supply lines at the end of the war; after all, several of my friends had died of starvation and cholera at the American-run P.O.W. camp at Bad Kreutznach, including Rolf Gast.

I had at that time a different calibration of what Germans stood for. I saw myself as belonging to a *Volk* of great thinkers and poets and musicians, with an honorable military history. I could not believe Germans had killed millions of innocent people. Only gradually, and only after the German government declared it unlawful to deny the Holocaust, did the whole truth sink in.

And only after I had lived years in the United States, where I often encountered and came to know Jews—people who treated my wounds and illnesses, gave me legal advice, did business with me; people whom I loved to watch perform, whose words I heard, read, and listened to, people whose counterparts had been physicians, lawyers, bankers, store owners, actors, musicians, conductors, composers, and writers in pre-Nazi Germany—did I realize fully what incredibly foolish, unspeakable, and unthinkably devilish leaps the Nazis had made when they first forced the Jews to leave, then picked up whole families and put them in camps, and finally destroyed them.

During the entire Nazi period, I never heard anybody voice any protest against the people in power. In fact, Hitler's government gave Germans something we valued: he returned our pride to us. He had flouted the degrading Treaty of Versailles imposed on Germany after World War I by the victors. Everybody was employed, and we could boast of our new *autobahns,* and even more of the power of the new German *Wehrmacht.* None of the people I knew said anything against this spirited, well-equipped force. The officers and men looked fantastic: they could goose-step as smartly and stand as erect as Prussian soldiers had. The Wehrmacht was made up of people like us, led by our best, mostly from the upper class. Its select social composition contributed to the extreme pride felt by its members, and also to the pride we, the German public, took in having such an admirable military organization represent us and protect us. The pride it gave us in being German differed from our pride in the achievements and notable works of the many great German scientists, composers, and poets: owning such a great instrument of power gave us all a feeling of great security and superiority.

Even we children admired our army. We began to

take a great interest in military equipment and training. When I was eight and nine, curiosity often led me to follow the Jungvolk group my brother had joined. I observed their play and actions in the forest and often listened in when they were being taught songs or facts about the Fatherland and Hitler. I liked the simple sounds the boys of the Jungvolk and the Hitler Youth could make with their shiny bugles and the way they beat their drums skillfully in rhythms. My brother's group thought it funny that I, the little kid, stayed with them but they accepted me as a sort of appendage or mascot. I found all their actions fascinating, and was amazed that a uniform could give a person authority over others. Just one element, a *kordel* (shoulder knot) or a special hat, could give someone a new identity: through that element, one became a representative, speaker, and enforcer for a greater power.

This fascination was the only one I ever had with a uniformed organization. Unlike my brother, who cherished uniforms and advanced quickly to leadership positions in the Jungvolk, I had no instinctive love for things military. When I had to join the Jungvolk at age ten, I was reluctant to give up control of my free time. My vague dislike for the power and authority of a uniform stayed with me through the war. As I grew older, I recognized even more clearly how a man in uniform, exercising the power given him by the state, would ultimately be willing to sacrifice anything and anyone for it. Even his life.

About that time, when I was eight and my brother was twelve, an adventure bound us closer together. It started on one of the outings with the Jungvolk, when Frieder and I discovered a lone little house in the forest. It was boarded up, and apparently no one lived there. We became very curious about it and wanted to explore it, even if doing so meant breaking in. My brother talked me into taking all my savings to buy an pistol. He told me we needed it, because otherwi breaking in would be too dangerous. Over a long time I had saved up seven Deutschmarks, mostly by selling my brother, who could never resist sweets, my desserts.

Air pistol at the ready, we climbed onto the roof and through an opening down into the house. Our excitement quickly dissipated. There was nothing to see. We left the same way we had entered without taking anything.

Even though my brother had told me not to, I made the mistake of telling my mother about this excursion. She became extremely angry—astonished and disturbed that we were capable of such an act of criminality, and appalled that we had purchased a real weapon, a pistol, the most impossible thing for anyone in our family to own or even think of after our father's death. She made us get rid of it immediately. I had to exchange it for something completely harmless. This is how I lost all the money I had carefully accumulated and acquired a flashlight I did not need.

Another attempt to satisfy curiosity began one day when a group of us came up with a question: how did girls pee without something we boys had and where did the pee come out? I had no sister and had never seen my mother nude. I did not know how girls were able to do this. We boys decided to investigate. We talked one of the little girls into going with us to an area where big bushes provided good cover. There she took off her panties, crouched down, and peed on the ground while we watched from below. I was disappointed that I couldn't see anything special. Unfortunately, the little girl told her mother what had happened, and her mother told my grandmother, and she in turn told my mother. I was scolded by her and perhaps whacked a bit. The adults saw this incident as a sexual transgression, something of which we boys were really innocent.

We continued our scientific investigations a little later. When questions about sexuality started to plague us, we started to play "doctor." As it turned out, nothing sinful ever resulted from our curiosity, and the distance between the sexes actually increased for us until we were about fifteen or sixteen. Germany in the 1930s was very different from the United States in the 1990s. Different interests, strict parents, no media influence, no sex education, no coed schools, and deeply ingrained Victorian attitudes and customs created barriers that were hard to break through before late adolescence.

When I was twelve years old, I did plague my mother by asking again and again how children are made. She was terribly embarrassed and never gave me an answer. Even though I had correlated cause and effect when my pet white mice, who frequently jumped on each other, had babies every six weeks, I could not imagine humans performing such an act. My friends and I were naive and innocent. Today's kids are taught in school all about the mechanics of sex. No one closes the projection-booth shutter for them when there are sexy scenes in movies or on television. For them, sex will never become the mysterious, secret, holy act it would eventually become for me, one where passion, affection, and honesty always played a big part.

1 They were discussing the 1935 Nuremburg laws—the first of the Nazi racial purity decrees. Further laws came in 1938. Columbia History of the World, ed. John A. Garrety and Peter Gay (Columbia University Press, 1972), 1055

4

Just across the street from my grandmother's lived my best friend, Karl-Heinz Blumenstein. Not only did we share the same first name: we did everything together. When we were about seven, we went off together for our first day in school. On this occasion both of us received the obligatory *Zuckertüte,* a three-foot-tall cone filled with candy and treats which were supposed to sweeten the loss of play time.

Karl-Heinz Blumenstein was even smaller and skinnier than I was. His wire-rimmed eyeglasses never fit very well, and their elongated oval lenses made him look comical. Blumenstein first attracted my admiration because he could go to the bathroom without taking his pants down, right in front of us kids on the street. He would first press out a piece of excrement; then, squeezing his rear end, he would let it fall down along his leg through the opening of his pants and over his shoes onto the street. His talent meant that he never had to leave the action to go up to the toilet. It was a remarkable skill, and everyone was fascinated, but as far as I know, no one tried to imitate him. Even he, with his uncanny ability, had to assess carefully the critical consistency his act required, or all could have ended in disaster.

The parents of most of the kids in the neighborhood soon decided that I was a wild sort of kid, an unsuitable playmate for their children. They knew instinctively that I could lead their kids into situations one could not anticipate and that I would make them wilder and harder to control. Parents in our immediate neighborhood began restricting their children's time with me, as my mother had when I had wanted to play with Bruno. Once I took all the kids on one of my regular explorations of the forests nearby; we got lost and had to spend the night sleeping on the ground. After that, none of the other kids was allowed to do much with me.

But Blumenstein's parents liked me very much, and used me as an example to inspire Karl-Heinz to apply his talents. When his father tried to teach us to play the mandolin, I took it and right away could play a piece of a tune. As I played *tata, tata, tatata, tata, tatata, tatatata, tatatata,* Herr Blumenstein told his son, "See, just do what Heinzi does."

Of course, neither of us could be the other. He was my friend because I liked him. I think he was intrigued by the ideas I came up with. Whenever I figured out the next thing to do or the next place to go, Blumenstein instantly became an integral part of the adventure, regardless of whether it was likely to turn out good or bad. He would get very excited, and would give himself a sort of handshake by twisting his hands together, and then drop down into a half-crouched position and press his hands and himself so hard that his face would become contorted into something resembling a badly squeezed red balloon with glasses.

At some point after Blumenstein had become my friend, my mother developed the belief, probably based on their family name, that his family was Jewish, or part Jewish. This idea did not affect our friendship, and we never asked any questions that might prove it true or false: that wasn't a question to ask. My mother worried, though, that the Blumensteins might have trouble. I never really knew: in fact, while writing this book, I called Karl-Heinz's younger sister Jutta and asked. She told me that they certainly weren't Jewish.

Blumenstein and I went to school every day. We kids had little perspective about why we would need an education: we went to school simply because it was required. Perhaps children whose parents were not of the working class had more of an idea of the purpose of education. We didn't, and had no desire to learn. But the teachers made us do so.

We didn't, however, refuse to learn. It was, in a way, honorable to go to school and make it to the next grade. Only kids from families that did not care or could not manage were held back a grade.

Our first teacher was Fraulein Lommer, a not very attractive middle-aged unmarried woman. We called her *Lommer Tante* (Aunt Lommer). She taught us the alphabet and how to read and write. She was usually very friendly and nice, but had the nasty habit of hitting us across our fingertip with a flexible ruler for the smallest infraction. We liked and disliked her at the same time. She taught us for three years, until she committed suicide. We were told she did it because she was lonely.

After her, we had Herr Mühler, a veteran of World War I. We learned about geography and history from him. I remember learning about the Germanic tribes and the Middle Ages. We liked him very much because when he got to telling us about World War I, he focused on the aspect of it that was most important to him—his own experience of almost dying in battle. We liked his story so much that we had him tell it again and again until we knew his story even better than he did, and could prompt him on every detail. Even though his eyewitness account described many people getting wounded, suffering, and dying, we felt excited and somehow proud to hear it, as if it were a story we had learned through personal involvement rather than memorization, as if we ourselves had been in the war. All of us were a little envious: we would never experience anything as exciting as that war.

Our next teacher was Herr Breitung, who was on leave from his duty as a U-boat captain. I remember the day when my class got new watercolor sets and some paper from a charity organization. Herr Breitung immediately gave us an art assignment: paint a dog. When I gave him my painting the next day, he asked, "Have you had help in painting this picture?"

I replied, "No, I did not." And Herr Breitung, thinking I was lying, gave me a big slap on my cheek.

Later, my mother and I ran into Herr Breitung as we walked over to the Christmas market to select a tree, and he apologized to my mother. My painting, which showed a dog as the companion and friend to humankind, had been implausibly good. He even told my mother that she should send me off to a better school since he believed I had potential and talent. Because of our always precarious financial situation, though, this was not in the cards.

I was never really angry at Captain Breitung. From his slap I learned that the world is not fair, and his praise at the Christmas market opened for me a small crack of confidence that injustices might sometimes be rectified. I was also grateful to him for correcting my poor posture and telling me to take my hands out of my pockets, even though he did this by mockingly imitating me—in front of the class—walking in the city, head down, with my hands buried deep in my pants. Not long after he taught us, Captain Breitung died when his U-boat was sunk at sea.

I knew by then that there were several things I did well at school. I was always the one who was most skillful working with his hands. I thought that my paintings were somehow the best. (I think now that they were naive but showed imagination.) And even though I was a very poor speller, I wrote the best essays. I was also capable in mathematics, particularly in geometry. I never thought of myself as one of the bad kids at school. Kids who were bad did no homework, or hit the teacher back if he dared to hit them, or kicked the teacher in the shins. They were from a lower class, even poorer than I and not well raised.

Even though the kids in my school were all from working-class families, there were class distinctions I sensed among us. While those bad kids were below me, others, such as Wolfgang Poppe, belonged to a higher class. Wolfgang's parents owned the little dairy and grocery store across the street from us. In our eyes his parents were rich people. They owned their house and had a three-wheeled vehicle to pick up milk cans and other goods. Herr Poppe's vehicle sometimes tipped over, and when that happened we would help him lift it back to an upright position. Wolfgang often helped out in the store and behaved, even when he was with us, with the authority of a proprietor. Later, just before the war, the Poppes even managed to buy a car, an old Adler. They were the only ones in our neighborhood who had an automobile. Immediately after the war started, though, their car was confiscated; so were all the work horses which at that time were still kept in stalls in the backs of some houses in our neighborhood.

Wolfgang had something that made him, to me, very different and unapproachable—a planned future. He would become the proprietor of his father's store. For that reason, we never became real friends or even comrades.

Though the Poppes were our social superiors, Wolfgang's parents made the same error concerning his education that my mother made concerning ours—not sending us off after the fourth grade to the *Lyceum* or *Oberealschule,* where we could have obtained a baccalaureate degree after another eight years of school. For the rest of our lives, we would have much greater choice in what to study and what career to pursue. Even Wolfgang's success-oriented parents could not see the future while working their little gold mine. They thought their house and store would assure their son's future. But after the war, when the Communists had taken control, the East German government classed the family as petty bourgeois, crippled their business with regulations that state-owned enterprises did not have to obey, bankrupted them, and made their store into a state-owned entity. Their house was confiscated for one Deutsche Mark. When Wolfgang's parents died, they left him an inheritance totaling only $1,200.

School we had to do; the things in life that absorbed us took place in the world outside. After school, Karl-Heinz and I roamed around and raided gardens for berries and fruits, carrots and kohlrabi. We never stepped onto anyone's property and only stole what we could reach through the fence slats. We ate everything right away. One of the gardens we raided belonged to Blumenstein's grandfather. Once we broke into the little garden hut there: it was so mysterious to us we had to inspect it. To shock the old man, we tied up a live mouse so that it was sticking out over the padlock, looking him in the eye when he came to open it.

We built countless play castles in the forest and beside the path to it, and saw to their maintenance. There we smoked a few cigarettes, quitting when it did not taste good. We also smoked the dry branches of a wild vine called *Rabing*, through the tiny capillaries of which we could suck the smoke. These branches glowed nicely, had a sweet taste, and were good for puffing thick smoke rings into the air.

Once while we were playing with matches, we accidentally started a grass fire in a meadow that bordered on a very dry pine forest. The fire started spreading at an unbelievable speed. Scared to death, Blumenstein and I took off our jackets and began pounding the flames like mad. We barely managed to put it out, preventing a major forest fire. He and I guarded the ashes for a long time to make sure the fire would not start up again. It was a major lesson for us, and we swore never to do that again.

Blumenstein and I came to know every spot in a wide territory beyond our neighborhood. We knew where to find wild strawberry patches, raspberry bushes, and blackberry bushes, and how to recognize the many rabbit, hamster, and mouse burrows in the wheat fields. Sometimes we dug for mice and found nests with young mice in all stages, from newborn to almost adult. A big labyrinth of such burrows could yield as many as a dozen adult mice.

Digging up hamsters was much more difficult. Often we would dig all day deep into the ground, while the hamsters continually evaded us, stuffing up their holes behind them. Digging for hamsters was very profitable, though, because as we dug we collected all the wheat they had stolen from Herr Coper, the farmer who owned the fields. Sometimes this amounted to a whole bucket of wheat, which we sold to people who raised rabbits in their backyards. Also, Blumenstein and I skinned the hamsters we caught, stretched and dried their skins on little boards, and sold them to a fur trader for one Deutsche Mark per skin.

We made money also by collecting "horse apples." Lots of horse-drawn wagons loaded with ice and big beer barrels passed by our house. We admired the huge Belgian horses that pulled them, but we were really more interested in their droppings. Blumenstein's Uncle Chrispin needed the manure for the little tomato garden in back of his house. We also picked up coal that fell off trucks, because Uncle Crispin needed it to heat his barbershop. We received five pfennigs a bucket for horse apples and the same for a bucket of coal.

Whenever we made a sale, we immediately ran to a

tiny store just around the corner from where we lived. The store was owned by an old woman named Haumersen. She had very few customers because she was a curious-looking woman and her store was in an odd location. We kids loved the store, because for three pfennigs we got a mix of *gummi men,* a licorice roll, and some bonbons. Our favorite candy, though, was *salmiak-pastillen.* These were black and diamond-shaped, and we got a large handful of them for very little money. Curiously, we called them *neger mesen* (black pussies). We were always tempted to order them from old Frau Haumersen using this secret indecent name, but never dared.

Eventually Blumenstein and I were joined in our play by a group of other boys of the same age. We had no playgrounds that were invented by adults; we created our own in the world around us. We knew, for instance, all the places where water would collect after a storm—new building sites with large excavations for basements and streets where architects and street builders had failed to create proper drainage. We built rafts from materials that the storm runoff had brought down the mountain, or from wood we found nearby, and played Columbus or pirates. We would sink the rafts of others and have our rafts sunk in turn, going home with our clothes dirty from these baths in the muddy waters.

We played war by digging big ditches and building bunkers in abandoned gardens. We made all kinds of weapons, including slingshots and bows and arrows, and even used BB guns to fight the enemy, the nasty children from the public housing project. Miraculously, no one got seriously hurt, but we had some ugly fist fights and wrestling matches. Many times our clothes were torn to pieces. We loved our improvised playgrounds, rough play and all, and we would not have ever exchanged them for the sanitized and sterile playgrounds where today's kids play.

Every so often a fad seized us. One year, every kid had to have a *Radelrad,* a wheel with a nail in the center which we rolled in front of us with a stick. You did not go anywhere without it. Later, everybody had to have tops—we would spin these on the sidewalk for hours. Then we all had to have bows and arrows, which we made ourselves out of hazel shoots. After the bows and arrows, we had a fascination with blowguns. A nearby glassblower made us long pieces of glass tubing with mouthpieces like a trumpet's. We used peas and all sorts of berries as projectiles, and we could shoot these some distance with accuracy. Every kid had a knife, some string, some marbles, a handkerchief, a comb, and some ammunition in his pockets. One time I tried to shoot off a man's top hat—he must have been returning from a funeral. I aimed from ten yards away and hit him right on his cheek. He was shocked, so it must have stung. If he had seen me do it, he certainly would have beaten me, but I had hidden the blowpipe instantly after firing the shot, and I got away with it.

Each season brought us new opportunities for fun. In winter, it was a great moment when the first snow began to fall. We watched every snowflake coming down, waiting for enough to stick so we could go out sledding on one of the steeply sloping streets not far from us—the *Forstweg* or the *Birnenstil* or *Kronenfeld Strasse.* I usually had a cheap sled that my mother bought at Woolworth's for about one Deutsche Mark. We called those cheap sleds *"kaese hitchen"* (cheese board things) because they were cheaply built, like the boxes used to ship cheese and fruits. They were slow, and if you did manage to build up some speed and then hit a dent in the road, they would break apart. But the way we used to sled could break even better sleds. We would start out at the transformer house high up on a little hill, build up speed zooming down the hill, and then jump a four-foot wall to continue down the street. Most of the accidents happened right there. It was the most exciting thing to do and to watch.

We also used to tie together several sleds, using the first one as the pilot car. With many kids sitting on this train, the pilot would guide the segmented vehicle down that same slope, over the wall and onto the street, and then all the way down the sloping street until we were close to the place where the bad kids lived.

We never thought about getting hurt when we went shooting down that hill. We never saw the trees on the sides of the streets. We did not think of accidents even when we mounted skates under boxes. By then the snow had melted and frozen several times and formed an icy surface, and we achieved truly incredible speeds on a course through the nearby forest called *die Hölle* (Hell). Eventually one boy was killed and several others badly injured. For all of us, this accident was a very sad and sobering experience, and we gave up this activity.

We also loved to skate—not on boring ice rinks, but on the Saale River or on sloping streets where trucks and buses packed the snow down till it became almost as hard as river ice. We would race down these streets at great speed; then, after arriving at the bottom, we would wait for slow-moving trucks or buses we could latch onto for a free ride back up the hill. The drivers hated us for doing this and would often stop to chase us away. Nevertheless, we usually managed to get a ride back up the hill. Skating was great fun, but it caused damage to our precious shoes: we had to fasten the skates onto them, and sometimes they ripped off the soles.

On one very cold day I broke through the ice on the Saale while skating. I was able to hold onto the ice sheet and was rescued by other kids who held out to me long branches from nearby trees. Not realizing how close to death I had been, I only worried that my wet clothes would freeze stiff and I would catch a cold.

We also liked to skate on Baker's Hill, where steps and steep slopes intermingled. We practiced jumping long distances, sometimes twenty feet or more. At first, the baker found it amusing, and everybody was entertained by our feats of daring. Then one day Harold Zenner, one of the best skaters among us, had a bad fall and broke his leg. After that incident, the baker regularly spread ashes over this marvelous practice run.

In summer my friends and I would often go to the *Schleichersee*, an artificial lake where I learned to swim. Even in April, when the water was still freezing cold, we would jump right in. One year I became quite sick with joint rheumatism, with much pain in my joints. On a day early that spring when I tried to swim one hundred meters, I almost drowned. At the seventy-five meter mark, I suddenly could not move my limbs, or even yell for help. I sank. At least one person must have seen me go under, because I regained consciousness lying on my back, being pumped back to life.

We also had our rituals. Whenever one of our pets died, we carried it up a hill and buried it close to the last garden near the forest. This garden belonged to a man who looked like death: we called him Totenkopf (death's head, or skull). His garden was so overgrown it resembled the hedge of thorns and vines that surrounded Sleeping Beauty. After our funeral ceremonies, we would call out loud, "Totenkopf, Totenkopf, Totenkopf!" and run away. Nothing happened. Eventually, though, he must have made the connection, because one day, without our having called out his name, he came running toward us, wielding a big ax, and chased us for more than a mile. We never went back to that garden.

At the time I did not realize that I was the instigator of most activities. After I had an idea, Blumenstein would go into his happy grimacing and contortions, and the other kids would follow along with us. It didn't matter whether it was building big dams across the entire street to contain the water when the snow melted, or figuring out a way to repair them quickly with big pre-packed rolls of snow after a truck drove through them, or making big slingshots using rubber hoses from the doctor's office, or making blowpipes out of glass of the right size to fit berry ammunition from a certain bush, or fabricating big black-powder firecrackers to use to split rocks apart, or making knives with deer feet for handles and using these to make bows and arrows, or fastening skates on snow sleds to make them incredibly fast: first Blumenstein and then the rest would follow.

The things I did sometimes pleased and sometimes scared my grandmother. She would refer to me as *"das Kroet"* (toad), her favorite term for an unusual person or somebody she did not know what to do with. She found it funny when I left my sweater out in the field to keep young, furless hares warm, but was not amused when I stayed out in the forest during thunderstorms. I loved to be out there when rain came pouring down; the world smelled so fresh, and I could watch gigantic thunder and lightning shows no human could create. I enjoyed seeing the water rushing down the hills and would sometimes go to a well-hidden pond to watch it flood and see whether I could catch a frog or find a salamander for my aquarium.

When I was in the third grade, my mother, who had by then been working for several years, solved the problem of my constantly torn pants. She made a long-term investment by buying my brother and me each *Lederhosen* for Christmas. For me she bought the biggest pair of pants anybody had ever seen a kid wear. They were so big that I was able to wear them until I was thirteen years old. By that time I had to mend them daily because the seams were ready to burst at any moment, but I thought they looked great and just could not give them up. Those wonderfully shiny pants were the envy of all the other kids. They had the whole history of my later childhood smeared onto them: prune juice and greasy hand stains from every meal (we did not use napkins in our house), as well as stains and scratches from sliding and rolling down every imaginable hill, tree, or meadow. One could never have created a painting as expressive as the one built up on my pants. Parting with them meant letting go of a lot of fondly remembered experiences—like a newly married man throwing away his bachelor address book. Years later my pride in these pants was validated when a friend remarked how much he had envied my aged, adventure-scarred *lederhosen*.

My most exciting discovery at this time in my life was that one could explode beer bottles by loading them up with carbide. In fact, I went a little crazy. The reaction worked best when the bottle held one-third carbide, one-third water, and one-third air. The beer bottles at that time were perfect for this purpose because they had patented closure mechanisms that held the plug locked in place while the gas generated in the bottles. After about three minutes, the bottle would explode into a thousand pieces with a really loud bang. Exploding these seemed perfectly safe because there was always time to walk away and watch from a safe distance, or to hide in a ditch or behind a tree. One time, though, when we could not find any water, we put carbide into a bottle and peed in it. We waited. The bottle did not explode. We waited longer. We looked at each other, wondering what we should do next. We could not leave the bottle there—it might hurt some passerby. Finally, we decided it would never explode since we had not used a full measure of carbide and there had not been much pee in the bottle either. I picked it up and tried to open it, but I could not, no matter how hard I tried. The pressure inside the bottle kept the closer fixed in place on the cap. So I gave it to Harold Zenner (whose broken leg had only just healed) and asked him to open the bottle. We had seen him open beer bottles by hitting the closer with his flat hand, popping them open as a karate master might break a piece of wood. I had just passed the bottle to him when it exploded with an earth-shattering bang. I was not hurt, but everybody else got glass splinters in the face and hands. Harold was bleeding like a pig. His hands were okay, but the bottom part of the bottle had opened his leg from his knee to his shoe in one very deep gash that went down to the bone. We immediately tied up his upper leg with my shirt and took him to the hospital. Things could have been worse: all of us could have lost our eyes.

We did not have the slightest inkling that war was about to close the curtain on our play; the bang of this bottle and the injuries suffered were a foreshadowing of the terrible times to come.

5

Since I came to America, I have often been asked what I did during the war. I have never quite known what to answer, because I have sense that when people asked me this, they are really wanting answers to more pointed questions: Did I sabotage war equipment? Did I kill Americans? Was I a Nazi? Did I gas the Jews? A German sometimes finds himself in the seat of the accused. I have to answer that first question in the same way a Holocaust survivor might: I tried to survive.

During the years 1939 to 1945, war was, not only for the Germans, but for almost every European, an everyday reality that colored the canvas of one's life with a heavy gray undercoating, permitting no one the opportunity to paint pleasant pictures of the present or future. For most civilian Americans, the war was happening in some other corner of the world. In Germany, the war affected every aspect of life, every thought; one lived it all day long, every day. When Hitler invaded Poland, the lights literally went out: no light could be permitted to leak out to the street, where even lights on cars and bicycles had to be dimmed to one small slit. In Germany, from that moment on, life itself was dimmed and obscured. And so was information from the world beyond.

Instead, all Germans were being exposed to incredible amounts of Nazi propaganda, misinformation of a kind and quality that had never been seen before. Newsreels and radio told us about Polish aggression against German civilians in Poland, and gave us the impression that the whole world was persecuting our country. Billboards everywhere warned us: *"Feind hoert mit!"* ("The enemy is listening in!")

My tin soldiers were now alive: frequently they were marching on the streets. In truth, I was at that time proud to be a German, and not of another nationality or race. There was something terribly German about the German army, something one could identify with, like a defining moment in a nation's history. I was proud we had only brave and fearless soldiers.

Food was rationed from the very beginning, and all merchandise swiftly disappeared from the stores. At school we kids were given Vitamin C tablets every day because we had no bananas, oranges, or lemons to eat. Other fruits were also very scarce. Before the war we used to get a meal every day from a Quaker organization, but that, too, stopped with the outbreak of hostilities.

People converted the basements of houses into air-raid shelters by installing iron doors and piling sandbags in front of their little windows. Everybody had to pick up and try out a "people's gas mask" instead of the promised people's car, the Volkswagen. Everybody had to practice going quickly to the air raid shelter whenever the probe alarm sounded. Its ominously nasty sound ripped through the peaceful day or night, penetrating down to one's bones.

After reaching the age of ten, all German boys had to join the *Jungvolk,* the Nazi version of the Boy Scouts. I reached that age in February, 1939. Blumenstein joined with me, and we were together in the *Jungvolk* until nearly the end of the war. Once the war began, it was made obligatory to give two afternoons a week to learning about the National Socialist movement, and to practicing marching in formation and giving the new German salute. How would you like to greet your grocer or milkman with a hearty *"Heil, Hitler!"*? It seems ludicrous now, but almost all Germans did exactly that everywhere they went.

The *Jungvolk* slowly converted my playgrounds—the fields where I had dug for mice and hamsters, the forests and meadows—into training grounds where we were taught how to read maps and how to surprise the enemy, usually *Jungvolk* kids from another part of town. To win, we had to attack them, wrestle them down onto their backs, and steal their *Wimpel,* a little flag every unit carried with it. We were being taught skills and attitudes that would make us ready to give our lives for the Fatherland and more specifically, for the *Führer.*

Only later did I realize how impressionable, pliable, and totally naive one can be at that age. For example, I remember seeing *Fighting for Honor,* a movie about a regiment of young cadets at the time of Frederick the Great. Half-trained, they were thrown into battle where they fought bravely and died heroically, marching in formation. For some reason, their heroic-looking discipline and death in an attractive uniform made me envy them.

In real life, however, I did not want to learn about the *Führer;* I did not want to give up playing my own games two afternoons every week. But when I did not show up one day and instead went swimming with my friend Manfred Süss, the whole *Jungvolk* group of forty people showed up and caught us. And believe it or not, they made us bend over, with two big guys holding our heads between their knees, and receive a whack on the buttocks from each of the forty other kids in uniform. They would also have made my disgrace visible by taking away my shoulder strap, but I did not have one. My uniform was always incomplete: I had only a brown shirt, which I wore with my Bavarian-style *lederhosen.*

After that incident, my mother had a talk with me and made me aware of my duties. She never believed in the

Nazi movement, but to this day I can hear her telling me, "If you make too much trouble, they bring you to Buchenwald; that is the place where they took all the Communists and criminals." My mother knew of that place directly, because she had married a man who was nearly taken there himself.

By this time my mother had given birth to my half-sister, Gudrun. I had not attended my mother's wedding and I did not take any notice of my young sister for many years. My mother, brother, sister, and stepfather lived a life apart. My world centered around my grandmother. I knew, though, that I could not ignore my mother's warning.

The *Jungvolk* leaders had us march wherever we went—through the meadows, through the forests, through the town. Sometimes we would march by the little fountain where I had played as a kid, and would attack, right there in front of all the appalled bystanders, another *Jungvolk* group, jumping everybody, knocking them down into the dirt of the street, and taking their shoulder straps away to dishonor them.

Then we would march on, singing songs with such lyrics as "When the golden evening sun sent its last rays, a regiment of Hitler's came to town." We had no idea then how appropriate those verses were in another sense, or that the lines of another song, "Exhausted gray columns walk in the sun through heather and sand" could have been composed for the German armies in their coming defeat.

When the war began, I and my classmates had just begun the fifth grade, with a new teacher. When he came into the classroom for the first time, we were all stunned: he had the most horribly disfigured face I had ever seen. When he introduced himself, saying, "My name is Engwicht," his mouth and fractured lower jaw made an odd movement and the whole class started laughing.

He exploded like a grenade and his face twisted into a threatening grimace. Suddenly there was dead silence. He told us how he had received his terrible injuries during World War I. We never laughed at him again. Each of us had, from then on, an idea of how terrible war must be, the same kind of war that we were now involved in, the same kind of war that our leaders were glorifying. Engwicht turned out to be a terrific mathematics teacher, and I learned a great deal from him.

Gundlach, our physical education teacher, was a member of the Nazi party. Each time we came to him to be trained to be fit and flexible, we lined up facing him, and he would stretch out his arm to greet us with the Nazi salute, which we then returned. He would tell us to remember our brave soldiers at the front. Then the calisthenics could begin. Since I was under doctor's orders not to participate in sports because of my heart, I could only watch what everybody else was doing.

Gundlach also taught us a great deal about plants and animals. We went on many field trips with him. He knew all kinds of mushrooms and loved to collect them with us. I learned all I could about mushrooms and started to collect them in my spare time. Our family ate a lot ourselves and I sold some to others. I stopped this activity only after I poisoned another family badly.

I remember vividly the building in which Gundlach lived with some other teachers from our school. I passed it every day on my way to school; at a time when billboards were rare, Gundlach's building had a big billboard on it, with a caricatured Jewish face and big letters saying:

"Trau keinem Fuchs auf gruener Heid,

Trau keinem Juden bei seinem Eid."
("Trust no fox on the green heath,
Trust no Jew's oath.")

Even though I was young and impressionable at that time, most of this stuff produced no positive response in me: I never have taken well to being told how to think.

After the war I saw Gundlach again: he was digging ditches. As a former Nazi in communist East Germany, he was never allowed to do any other kind of work again. Even today I often think of Gundlach, because every day I still do some of the gymnastics he taught us.

In the first year of the war we suffered through many air raid alarms. The ones at night were particularly annoying. They usually woke us up after we had settled into deep sleep. I hated being awakened in the middle of the night and then having to get out of bed and get dressed, walk down the three flights of stairs to the air raid shelter, and sit out the alarm crowded together with everybody else from the apartment house. The neighbor who bothered me most in the shelter was Frau Gross, a peculiar lady who wore, winter and summer, what appeared to be a big thick tea cozy pulled down over her head so that only a small part of her face could be seen. She said her head always felt cold; now I wonder if she might have been bald. She once placed an unusual ad in the newspaper: "Exchange one bra, size 40, for size 42." Another time, she flooded her entire apartment and also the apartment below her. From then on my grandmother and I called her *"Frau Hochwasser"*—Mrs. Highwater. But in English the name *Gross* was an apt characterization of her because she had a seemingly unlimited supply of gas in her belly and farted uncontrollably into our airtight shelter, apologizing each time.

When we heard the anti-aircraft guns for the first time, we were all surprised at their loudness and their peculiar metallic tone. We sensed that the shelter's walls, sandbags, and iron doors would provide little defense against a bomb or a crashing airplane.

After the first night of this activity, one boy brought to school some flak shrapnel he had found on his way. He lived the farthest from school and was famous for collecting rain worms in his pockets; he had once eaten one right before our eyes, for money. We were all very excited by the splinters. After school, we ran out to search for shrapnel. Over time I accumulated a whole collection of deadly bits of metal, even finding for myself a piece with a part of the copper ring from a grenade. It was especially valuable and precious. We had no idea that later on, carpets of much larger fragments from bombs and flak grenades would litter our streets.

Soon alarms became routine, and, we thought, many times false. My grandmother and I finally got so tired of going to the shelter for nothing that we decided to just get out of bed, stand at the window to observe any action, and talk to the neighbors in windows across the street. We would only go down to the shelter if it seemed dangerous to stay upstairs. While I was standing at the window one night, looking up into a beautiful, starlit sky, three huge planes, like giant condors, suddenly flew toward me and over the house. They seemed low enough to touch with my hand. Then I heard four or five loud, dark-sounding detonations. I fell to the floor—I'm not sure whether the blast knocked me over, or whether I fell over trying to get away from it. I had witnessed the first bombs that fell on Jena. We were all shaken but excited; we had made contact with the enemy, after all. When the "all clear" sounded, many people came out of their houses to look at the damage.

This bomb, the very first of our war in Jena, hit the *Horst Wessel Schule,* my school, and damaged the building quite badly. We kids were ecstatic: we could not go to school for several months. Two houses close to the school had also been hit and their side walls had been blown away. We could look into them and see how the people had lived, for most of the furnishings were still intact and standing in place. It reminded me of the puppet theater I had played with as a child, but it was not the same. Some of the people who had been living there had been killed, right here in our town.

These people were given a big funeral ceremony on the market square where I used to lead the band. We heard speeches that told us that they had died in the service of the Fatherland, and that we would somehow avenge them. The speeches sounded more like a celebration for later victory than funeral sermons. Still, these victims had been fortunate in one way: they were honored by the community. Later on, no one got an official funeral. Those who died were just loaded on trucks and carted away.

I don't remember anybody explaining why the planes had dropped bombs on our school, but it was not by accident that they made a run there. It was a failed but incredibly brave and daring attempt by those flyers to knock out a part of the Carl Zeiss Jena factory, 300 yards from our school, which manufactured U-boat periscopes.

For some reason this whole episode did not shake us too much. After all, we were at war; Poland, Belgium, Holland, and France had been overrun and many soldiers had died. Wasn't it natural that some of us should die, too?

By that time we had become accustomed to the exciting action shown in the newsreels: we admired the Messerschmidt ME-109s downing enemy planes, the Stukas bombing enemy troops and towns, and the tanks rolling through villages and even through houses, or knocking down trees in forests and then shooting to destroy enemy tanks. All of us were very proud of our soldiers and believed they fought with more heroism and bravery than their enemies. The newsreels showed to us common soldiers being awarded the Iron Cross first or second class by their staccato-speaking officers, who in turn were decorated with the higher *Eichen Laub und Schwertern* (Oak Leaf and Sword). We believed these officers and men with their handsome uniforms, poise, and *Opferbereitschaft* (devotion) were superior to English, French, Russian, or American soldiers.

Almost all of us had our maps of Europe on which we plotted the German advance, planting colored pinheads in each conquered city, following these gains as one might follow the prices of individual stocks today. We listened to the *Frontberichte* (front updates) routinely, like the daily market summary. We turned up the volume when the *Oberkommando der Wehrmacht* began playing Liszt's *Les Préludes* (reduced to a short jingle) to announce that a U-boat, perhaps one with a famous commander, had sunk another high-tonnage British-registered freighter. I was very fond of German war heroes, and particularly liked Captain Lieutenant Brien, the commander of the U-boat that sank the most ships, Oberst Gallant, the pilot who shot down the most enemy planes with his Messerschmidt ME-109, and the Desert Fox, General Erwin Rommel, the general who conquered large portions of North Africa.

In the reports of all these triumphs, loss of human life seemed to play a very minor role. When ships were sent to the bottom, the announcer gave a figure for the tonnage sunk, but did not number the crew members plunged into the Atlantic to drown. And if the *Oberkommando* did mention German losses, they were painful, but as long as ten times as many enemies had been killed and a battle won, they seemed acceptable.

While the damaged sections of our school were being sealed off with walls, we students were sent out on big campaigns to collect *alt material*—old scrap like copper pipes, iron, aluminum, and old cloth—to support the war effort. Blumenstein and I knew every place things like this could be found, and we became the best at collecting everything. But a question nagged at us as we collected this material, a question we could not answer: if the Fatherland needed all that worthless stuff, how could we ever win the war?

We also went out in the fields to collect *Heilkraüter,* rare plants for the production of medicines. We would start in the morning with a big four-wheeled cart and collect both firewood and those useful plants in forests and meadows far away from our house. At the end of the day we would pile everything behind our damaged school building. We lost a lot of school time because of the slow pace of the repairs and ended up with an even worse education.

Later in life I was frequently reminded that I had failed to learn, or my teachers had not made any effort to tell us, about the beauty and power of words, their ability to evoke feelings and scenes, their ten thousand years of roots and origins, their relationship to other languages, and the meanings of their many different prefixes and suffixes. I had not even been taught how to place commas properly, nor had I learned what genitive, dative, and accusative cases were.

I had, by this time, become conscious of the caste system in German education. At the end of fourth grade some of our classmates had gone on to other schools, the *Lyceum* and the *Oberealschule,* which would prepare them for academic studies and careers. During the next eight years they would learn modern foreign languages and Latin, explore the humanities, and progress through higher mathematics. Those who were successful would move on to another six years of intensive study at a university.

Those of us who remained behind in *der Grundschule,* on the other hand, would receive only four more years of schooling, intended to prepare us for our futures as apprentices in industry or as tradesmen—butchers, bakers, plumbers, and so on. We were destined to live small lives, to become the underdogs. Even in the *Jungvolk* our peer leaders were our academic superiors, students who were attending the *Lyceum* and the *Oberealschule.* They, not we, were to become the future officers of the German army.

When I finally realized, at age ten or eleven, that my schooling would determine my future, and not in any positive way, I was disturbed. I tried to think of a way to change my situation. I saw no chance of getting into a more academic school: my mother had let the application deadlines pass, probably in part because she was afraid of having to support me for a long time. I played with the idea of going to the Napoli, the strictest, most demanding educational institute for Nazi youth. But my long non-participation in sports would have prevented me from being accepted there, and besides I did not have the clean-blooded German appearance they were looking for. I think I also sensed that, with my attitudes, I would not have survived the regimentation.

There were no alternatives. I was stuck, and so I settled for learning a bit of French from the French prisoners of war who worked nearby digging ditches. Any contact with POWs was, of course, strictly forbidden, but Blumenstein and I loved to talk with them in sign language and bring them little scraps of food and fruits. The first word we learned was Blumenstein's name in French: "Fleurpierre," or flower stone. Later, I called him this when he was in trouble. Unfortunately, my name could not be translated into French, but I learned how to pronounce it with a French accent. Even today I remember particular words those warm and wonderful men taught me.

When we needed money in those years, we would walk to the tennis courts and pick up balls for the people who played there. It was very hot work sometimes, but paid relatively well. One kid in our neighborhood, a year or two older than I, saved up enough money from this work to buy a bicycle; he later died as a soldier.

After two years of war came 1941—a year no German would ever forget. Hitler invaded Russia. We first heard the news on the radio, and we were so shocked we wanted to run down to the street to share the news with others. We felt almost betrayed by this unfathomable undertaking, because at that point victory and peace seemed to be close. Other people had already gathered below the staircase on the ground floor. A refugee from Silesia who was living on the ground floor made the most ominous and courageous statement I heard from any private citizen during the war: "This is the end!" It hit me like a heavy stone on the chest, because I knew right then that it was true. I lived through the rest of the war, through all the horrible things I saw and had to experience, with that certainty.

After the commencement of the attack on Russia, conditions for civilians grew steadily worse. Food was rationed more severely. Each person was assigned a number of calories needed to work in his profession: laborers who shoveled coal and intellectuals were allocated the most; children and housewives received the least.

By this time our family at home consisted of only women and children. My Uncle Alfred had been drafted to serve on the western front in France; my Uncle Helmut was running a "front kino" (movie theater) in the Balkans; and my stepfather was on the eastern front.

My family, like others we knew in our working-class neighborhood, openly debated how best to achieve our objective—surviving unharmed. Not one of us harbored enthusiasm for the war or was eager to die for Glory and the Fatherland. Most of the soldiers we knew hoped to avoid battle. My uncles tried to think of ways to become prisoners of war without first getting wounded or killed by the enemy or by their own side.

My mother worried most about my brother Frieder, who had actually liked the Hitler youth organization, loved uniforms, and, with the help of his Aryan good looks, had been promoted to be a *Jungvolk* leader in Jena. My mother would never have allowed him to volunteer for the military, but if the war lasted another year, he would certainly be drafted. I remember my mother saying that now the only safe place to be was in a concentration camp with the political enemies of the state: there you wouldn't get drafted or bombed. Not until the war ended did we learn the truth about conditions there.

We watched with intense interest the advance of German troops far into Russia, but even the many announcements of victories did not convince us that this venture would end well. Our skepticism was soon reinforced when the hard Russian winter stopped the advance and froze the German armies in their tracks. The pictures of Napoleon's soldiers freezing to death in Russia came into our minds. Everyone on the home front was asked to give up any luxurious fur coats and warm clothing and send them to our soldiers.[1]

When the Russian spring brought a Russian offensive, we heard the ice cracking all the way back to Germany. It sounded as if Germany itself were breaking apart.

1 German planning had assumed that Russian resistance would collapse before the winter of 1941; accordingly, the leaders believed there would be no need to supply winter clothing and equipment. In the winter of 1941 a drive was hastily organized to collect winter clothing and equipment. (Weinberg, 480)

6

Our *Jungvolk* leaders kept telling us to drill harder. We had to be tougher; we had to learn how to shoot better; we had to run faster; we had to sing louder, ever louder. Our leaders kept demanding more of us.

One time we had to race for twelve miles over steep hills and valleys while pulling and pushing wagons full of bricks. Our marathon ended at a little village, where the bricks were used to build a shelter. Another time we marched with full military backpacks from Jena to the Saale Tal Sperre dam. With supplies for a two-week stay, the packs weighed more than sixty pounds each. We carried them more than sixty kilometers, or about 40 miles. We started walking at eight o'clock in the morning and arrived at our destination on the lake at eleven o'clock the next morning. This exercise gave us an idea of what soldiers had to go through when they marched. Most of us, because of the lousy shoes we wore, had walked big blisters onto our feet.

Fortunately, we were allowed to stay at the lake, and so our feet could heal. That respite turned out to be the best time I ever had during the war. The place was beautiful, with extensive forests and water. We built rafts and staged dramatic sea battles. We improvised a performance of Friedrich Schiller's play, *Wilhelm Tell,* and it turned out wonderfully. Schiller's work took heroism and patriotism very seriously, but the young man who took the role of Tell had been stung by a bee right under his eye, distorting his face so much he seemed to be wearing a carnival mask. Thanks to this comic touch, we ended up with a hilarious performance that gave us relief awhile from the hard realities of war.

At night we were often sent out to spy on and raid other *Jungvolk* groups. One night we walked several miles to attack a rival camp. I was always a little scared of what I might encounter and whom I would have to fight. That night, close to that group's camp, I encountered the biggest guy I had ever met. I jumped him, but the guy just would not go down. When he finally had me down and almost on my back, my hand pressed into something very soft. I realized my hand was in a big pile of shit someone had deposited there. I felt it going through my fingers and it was the most disgusting sensation I had ever had. But that handful of shit turned out to be an excellent weapon: when I threatened this big guy with it, he immediately volunteered to lie on his back. Sometimes misfortune works to one's advantage.

At this time, the German *Luftwaffe* was still strong; from our homes we could often watch truly spectacular air fights between our Messerschmidt ME-109s and enemy planes, usually big bombers. One day an American plane was hit, and we saw it go down close to Vollradisroda, a nearby village. We kids had been there years before, when I led the others into the forest, and we all got lost and had to sleep there. Since the village was not far, Blumenstein, Rolf Gast, and I ran to see the crashed plane. We ran for more than an hour, up the mountain and through the deep forest to the place where we figured the plane had gone down.

It was a gruesome sight. The plane had plunged into the big trees, leveling many of them before it came to a stop. There was no crater or fire; the crew had jettisoned their bombs before they reached the trees. One crewman had leaped from the plane as it approached the tree tops, but had been smashed on the branches. We saw nobody in the wreckage and were afraid something would explode at any moment, so we grabbed some rations, some chocolate, and a long string of heavy machine-gun ammunition, and ran off.

This was the longest and farthest I had ever run, probably five miles. Afterwards I realized that I was healthy and that I had recovered from my childhood illnesses, including the heart valve defect that had kept me from playing sports for so long.

Weeks later, after the authorities had done their job, we went back to the plane wreck. We dismantled many of the servo-motors that operated the flaps and ailerons of the plane and we also took some little metal shields that must have come from the bullet-proof vests the fliers had worn.

Afterwards, we made our own explosives from the ammunition by breaking the bullets out of the cartridges and then squeezing the powder-filled metal jackets closed with a large vise. We were very careful. We knew it was very dangerous, but we did it anyway. We saved these closed cartridges to throw into fires we lit outdoors, away from where any people lived; we would run away to a safe distance, and then watch them explode. Once when we did this, an old woman came out of nowhere. Seeing the fire, she walked towards it to see what was going on. Instead of calling to her, we all ran towards the fire to kick the cartridge out. We had just reached the fire when it went off with a deafening bang. I was not harmed, but one boy had some metal from the copper jacket penetrate his arm. Another, a boy named Schoeneweg, which, oddly enough, when pronounced in our dialect sounded like "beauty gone," was hit by a fragment that went in one cheek, clear through his mouth, and out the other cheek without touching a single tooth. He lived only two hundred yards from where this happened, and wasn't even bleeding much when we brought him to his mother, who was grateful that he was still alive.

When we began to be told of "front corrections" on the Eastern front, we understood that things were not going well there. We thought that the armies probably had to leave much of their equipment behind them. We boys could only imagine what might be left behind for the Russians who, like us, were collecting scrap.

The disappearance of many foods offered another indication that Germany was not winning this war. By this time, we had almost forgotten about *bratwurst,* whipping cream, and Danish pastries; they were things of the past. Wanting to know more about how the war was progressing, we started listening to the British Broadcasting Corporation. We kept the volume low and pressed our ears against the loudspeaker, so no one could hear what we were doing and report it. Hearing the enemy talk to us was incredible, especially since we suspected that the BBC's information might be truer than our own. Once we had heard the distinct four drum beats of the BBC, we could never forget it.

By now my brother Frieder had been drafted into a non-fighting unit, the *Reichsarbeitdienst* (State Labor Service), an organization Hitler had created originally to mobilize all young men of eighteen and nineteen to build the *Autobahns,* divided highways with two lanes in each direction. Now its members were building fake factories to draw attacks from Allied bombers.

In 1944 I reached a point where I had to make some decisions about my future. After eight years of school, we were about to graduate. At this time most of us entered religious studies classes in preparation for being confirmed—a major event for the parents. I seldom attended the classes. I liked some of the Bible stories, but I found the dogma that religion presented as truth impossible to believe in. Eventually I received confirmation anyway.

A state institution called the Office for Work gave

us new graduates a test made up mostly of "yes/no" questions, with a few longer questions. This test was supposed to discover our talents and our inclinations toward particular professions. I had been given no advice about what to do with my life, but I did at least have the example of my brother Frieder to guide me. He was just completing his apprenticeship as a *Feinmechaniker* (skilled technician) for fine optical and mechanical instrumentation at Carl Zeiss Jena. I had almost no idea what this work involved, but I liked the name *Feinmechaniker*. I also believed that a profession like this would get me at least a few steps higher on the ladder of status, and would provide me with a good foundation for a career.

I think I did have some awareness by then that I was someone who could invent things. In the sixth or seventh grade we were given lessons about the human body. When we came to the ear, I figured out how one might work. Before the teacher began to explain how the human ear works, I raised my hand to tell the class about my idea, and amazed my teacher and classmates with my description of a perfectly functioning hearing mechanism, which worked quite differently from how our ears do.

After finishing the apprenticeship test, though, I really had to laugh at myself. While the other applicants had sketched gears and gadgets to complete the drawings on the test, I had turned them into butterflies, animals, and plants. I was very worried that I had failed the test. I was accepted, however, for training at Zeiss, mainly because I had good grades in mathematics and geometry and had topped everybody in a sort of intelligence test that timed how quickly one could put hundreds of pins into holes. Having picked a lot of berries, I knew how to pick very fast, alternating hands while looking and feeling at the same time. This skill had saved me.

I made up my mind to give my apprenticeship at Zeiss the best I had. I dropped almost everything else. I did not even go out to the streets any more to talk to my secret girlfriend, Ruth Vordank. She was a beautiful girl my age with a cute freckled nose who probably never quite realized she was my girlfriend. I gave up reading the *"Colonial Hefte"* propaganda that informed us about everything in the former German colonies and how the English were bad but honorable, while the French were really lousy and dishonorable. I quit reading *Die Deutschen Heldensagen,* sagas of the German heroes of the distant past. I had read them so many times I knew them by heart, anyway. But I continued to read the books of Karl May, all 76 or so of them. May was a dreamer who wrote reports of his adventurous travels and adventure stories set in distant countries. His heroes Old Shatterhand and Winnetou lived in regions hardly reachable, even by the mind, in the Wild West of America.[1] May's works instilled in me a great desire to travel to faraway regions and to experience adventures there myself, even after I learned the truth about May: he had never actually traveled to any of the places he wrote about. I wanted to believe that he had been everywhere and was himself the hero he described, changing from Old Shatterhand in America, to Kara be Nemsi in the Middle East and Africa, and even into the Spion von Ordry as he adventured into different occupations, adventures, and spaces. It was too painful to think all these tales had been fabricated in an attic, not lived, and that May was not actually describing his own experiences, as I am now telling mine.

At Carl Zeiss, we apprentices worked the same hours as the regular workers—ten hours a day, six days a week. I began by learning how to scrub metal: at first I used a big scrub file to reduce a cylindrical steel block two inches long with a two-inch diameter into a rectangular one, and then continued with finer files to achieve smoothness, flatness, and very precise dimensions. Gradually we made ever more complex metal pieces in all kinds of complicated shapes and forms, and learned all the different metal-working procedures, such as forging, casting, hardening, soldering, and welding. We also learned how to use all the many different machines for working metal and plastic. In the last year and a half of the program we worked in different factory departments, learning to assemble and test various optical instruments as different as measuring microscopes and planetarium projectors.

Zeiss's apprenticeship program was the best of its kind in Germany, and I learned my trade well. I even scrubbed and polished challenging pieces for others who could not do the work properly themselves. Some of the apprentices came from the country, where food was more easily available. For one day's work, which I had to carve out of my own schedule by working harder, I would get a fantastic farmer's sandwich of homemade sausage, fat, and eggs on wonderful home-baked bread. I can taste them still, and they motivated me to become one of the very best apprentices in the program. Also driving me was the fact that my brother had been the best before me, and I was expected by the masters to become as good.

I got my last slap in the face in that workshop, when I absentmindedly played with a running transmission belt—a very dangerous thing to do. In the workshop we were always pulling practical jokes on each other. I remember sending a new kid to a master whom we secretly called *Arsch* ("ass") to pick up some tool. We always had to address our masters with *"Herr"*; this apprentice, not knowing the master's real name, addressed him as "Herr Arsch." Arsch leaped from behind his desk and gave the poor guy a beating, while we almost died laughing, hiding under our workbenches and machines.

With air raids a constant threat, all apprentices were trained to be firemen. Equipped with steel helmets and uniforms, we looked like little soldiers. The boys from the Lyceum were given the more important duty of serving up on the tower, manning four-barreled anti-aircraft guns. We envied them, even though during the air raids we could go to the shelters, while they had to sit up on the tower fully exposed to the attack.[2]

A lot of townspeople, many of whom I knew, came to the Zeiss shelters during raids because the reinforced concrete walls were far stronger than the stone or brick walls of homes. One day, after the alarm had sounded and many civilians had come in, I was standing near the opening of the shelter talking to a girl I liked. I was saying, "Why do you come here? This is really not safer than out there." As I was saying "really not safer," a bomb exploded, and I saw the opposite wall break into fiery cracks as it toppled toward me. That was the last thing I saw. When the dust cleared, people found me twenty-five yards down the aisle, unconscious but unhurt. After a few moments, I got to my feet and asked matter-of-factly what had happened. The others were in deep panic and shock. People had been killed, and the guy who had been standing opposite me on the other side of that entrance was buried and had lost his legs. Like me, the girl had been blown into the shelter and survived. The bomb had exploded outside, just fifteen yards from where I had been standing, and had blown everything in between into the aisle, including two walls and some transformer equipment.

Everyone else was in shock; my unconsciousness had protected me from realizing the full effects of the bomb. So I was the only one who volunteered to make a run to the Centrale to report our situation. I climbed over the rubble and out to the yard. The Zeiss facilities were extensive; I had to run about 400 yards. Some unexploded bombs were lying around. I stopped to inspect them because I had never seen a bomb. After I reached the Centrale and made my report, I rested; then everything I had seen and experienced began to sink in. I realized what could have happened to me. Now shock hit me, and I was trembling and fearful.

After some time, the people at Centrale went outside to take a look. Shortly afterwards I heard several bombs explode and thought at first it was another attack. Then I realized that these were the bombs that I had just climbed over. These time-delay bombs caused more deaths and injuries than the bombs that went off during the raid.

When we began assessing the damage, perhaps the worst we saw resulted from a direct hit on the building that fabricated submarine periscopes. A bomb had gone through two reinforced concrete ceilings to explode right in the air raid shelter. About fifty people had been inside, and their bodies, almost all of them dismembered, were hung and wedged among the broken concrete blocks and iron reinforcing rods. The sight was terrible. All of us went to work immediately, digging to see whether anybody was still alive. We dug with our bare hands, with iron bars and shovels, with pick axes and saws for days. After several days, we found three people alive, hidden under fallen locker cabinets. They were from Holland, foreign workers helping the war effort.

The damage was repaired quickly. Russian prisoners of war filled in the crater from the bomb that almost killed me.[3] They looked miserable and very foreign to us—they must have come from Mongolia or Siberia. We had seen newsreels in which tens of thousands of Russian soldiers, driven by their commissars, ran toward the German lines, only to be mowed down by the waiting German machine guns. We assumed these prisoners were the pitiful leftovers from this carnage. Sometimes from our tower we threw some of our lunch leftovers down to them, much as one would feed pigeons. They always ran for every crumb, which we found amusing. We knew it was forbidden to give them anything, but we would have given them more if we had not been so hungry ourselves. It would have made the scene less cruel. We could not imagine that soldiers like this would soon occupy German soil.

A later day set new standards of terror and made us realize that we had not been able to imagine all that was possible in war. The day started for me like most other days. I was at home with my grandmother. The sirens went off in the middle of the day, and all of us in our house crowded into the basement shelter, as usual. We were by now used to the flak bursting overhead, and to the many balloons above Jena dangling steel cables to protect our town and factories from enemy attack. We were standing under the arches of the shelter talking, when suddenly we heard a powerful noise like thousands of huge air blowers pushing air. It was right over our heads. It was so loud that it dominated our attention and partially drowned out the noise of the cannons firing. And then we heard hundreds of detonations. Our building and the ground under us had begun shaking; mortar and sand fell down on us from the ceiling. This incredible noise and concussion repeated and repeated, increasing in intensity, making us bend down as if we could support the ceiling with our backs should it collapse, until finally we hunched down close to the stone floor, arms and hands over our steel helmets. The attack lasted for a long, long time, as wave after wave of planes dropped bombs. Each second we felt that the very next bomb would hit us; each time a bomb hit, we thought it was so close that nothing could be left on the streets above us. Finally, it stopped. We waited in silence. We stayed frozen in position until the sirens signaled that it was over. We had survived.

All of us looked as if we had been shoveling cement and sand in a mine shaft. Haltingly, we walked up the stairs, expecting to see the building above us gone, with everything lost. Instead, everything was still standing. Even the neighboring houses looked undamaged. Where had all those bombs landed? The older people prevented me from running to find the damaged area. There were likely to be time-delay bombs on the streets and under the rubble. Soon we saw immense mushrooms of smoke rising from the old center of town and knew that it must have been destroyed.

During the attack I had thought only of myself and perhaps the other people in the shelter, but now I thought of my mother, brother, and sister, my aunt and cousins; they might all be dead. The thought that my grandmother and I might be the only survivors of our family pressed heavily on my gut. I wanted to go find out.

When I said that I wanted to run down the street into the inner town, the refugee who had told us that the Russian campaign would be the end said, "You have time to see all this damage, because you will have to work all your life to rebuild what was destroyed today in half an hour." He was so right. In Jena large areas bombed and incinerated during this attack remain ugly, improvised parking lots even today, fifty years after the war.

Nobody could prevent me from going, though. The air smelled of exploded powder and smoke. From afar I could see an incredibly big, blinding fire burning on top of the Zeiss buildings. The magnesium alloys from which we built binoculars had ignited. When I went closer, I could not believe my eyes. I had never imagined a fire as big as this. Whole city blocks, including areas I had roamed as a child, were either bombed into rubble or incinerated. Flames seemed to be reaching as high as our church tower. That tower itself had been hit on its top and was on fire. It was ridiculous to attempt to extinguish such a blaze, but some firemen were pointing their hoses at the flames anyway. I joined them, but it was no more effective an effort than if a village fire department with a hand pump had tried to save a large city from being consumed.

I set out to try to help evacuate buildings that were about to be consumed by the flames. The fire was spreading so fast that we had to get farther and farther away from it. Seeing that it was hopeless, I was tempted to quit helping and just watch the fire—I knew I would never in my life have another opportunity to see such a fire. I stopped to rest awhile in the big entrance of a department store, next door to the church. From there I watched the drug store across the street go up in flames. I saw the proprietor—the man who used to sell us dried sugar cane—run back in to save someone or something. After just seconds he came out burning like a big torch. I tried to extinguish the fire on him, but other people were faster in helping him, and carried him away. (He lived: I often saw him later, walking in the street. He had no hair, the skin of his face had burned down to the bone, and he had only two holes as a nose: he was a living representation of what had happened to Jena.)

I went back to my place at the store entrance. The houses across the street fueled a wall of fire three times higher than the buildings themselves and sent such incredible heat toward me that I had to cover my face. I also watched the church tower burn until the upper portion collapsed and fell in on itself. A few seconds later, burning debris and millions of sparks scattered down its outside and landed not too far from me. I was never afraid during all this and judged the dangers calmly, just as an umpire in a baseball game watches the field.

I saw a man going into the church and followed him for some reason. I think I wanted to see the damage the collapse of the church tower had caused inside. But it was calm and peaceful within, even then. I could see part of the sky through the roof. The man picked up some things and asked me to carry a big wooden angel to safety. This figure represented the Archangel Michael, the patron saint of Jena. I carried the statue out of the church and all the way over the Camusdorfer bridge, but it was so heavy I could not carry it any farther. I left it with the other man.

Almost exactly fifty years later, when I was in Jena visiting the Stadt Museum, I saw that figure on display. It was much smaller than in my memory; but I myself had been smaller back then, and my recollection of how heavy it felt when I carried it must have distorted its size in my memory.

Next I returned to the scene to help people who were trying to save their belongs. I ended up helping in a house that had a bakery in it. By this time it was early morning, and I was completely exhausted and famished. I went down into the bakery and took two loaves of bread. Hiding them as best I could, I slinked past people still working to save belongings and ran home. Even though that house had been condemned and was being left to the flames, I would have been shot as a looter had anyone caught me with the bread.

By early 1945 everyone knew the war was nearing its end. Almost all men older than I had been taken by the war: some were still fighting, some were now prisoners, and a great many had been killed. Germany was starving, and was being bombed relentlessly every night. There were now so many airplanes flying over Germany that I did not know where to hide anymore. I went a few times to a big concrete bunker with walls three yards thick, but it always felt like a mass grave. I went to a large tunnel dug hundreds of yards down into the sandstone for an underground factory. But when I was there, Allied planes bombed the north railroad station with huge bombs that made craters forty yards across. The earth shook so badly that the raw, unsealed sandstone and earth almost buried us alive. I went to the forest and got well shaken up there, too, when the bombers attacked, by accident, a nearby foreign workers' barracks. Nowhere was safe. Staying in the shelter at home meant certain death if a bomb landed nearby because of the thin walls; but the sturdier buildings at Zeiss where I was still working were more likely to be bombed. One time I stood on our street and counted more than 1,000 planes flying unhindered over our house. Their target was Dresden, where 80,000 people met death later that day.

One of the most horrible attacks happened near our railroad station in the Paradise Park in which I had played so peacefully. A long train had stopped there during an air raid alarm and hundreds of people had fled the train to find shelter under the trees. The attacking planes must have spotted the crowd; they dropped hundreds of shrapnel bombs into the park.

I arrived about ninety minutes later as part of a cleanup team. I don't think a single person had escaped—the density of the gouges on each tree in the park showed it would have been impossible. We loaded many bodies onto trucks, all with numerous holes in them. This was a scene I have never been able to clear from my mind. Even today, when I see a woman's white flesh under silk stockings, it can bring back this scene and the awful feelings I had when I had to grasp a young woman's body by the legs and throw it onto the back of that truck.

Just after that attack I was ordered to a camp to be trained to meet the coming American attack. Most males had already been drafted. When my brother Frieder had come home to work, he had been drafted for an elite unit of the *Fallschirm Jaeger,* called *Feldherrnhalle,* the name of the place where Horst Wessel was shot. He was being trained as a paratrooper. Now every remaining male from the ages of fifteen to seventy-five had to go to camp to learn the use of weapons.[4] None of us believed that this ridiculous-looking crowd of kids, old men, and leftovers could stop anybody.

We were going to be armed as tank-destroying groups. We would face the enemy within days. They gave us real rifles—but there was no ammunition. We were trained to take apart and reassemble our rifles, but no one ever fired a shot in the ten days we trained. We knew a rifle wouldn't even annoy a tank, but they promised we would have bazookas. They gave us bazookas—made of wood. The army had no real bazookas to spare for training. We practiced lying on the ground and hiding behind trees at the edge of a forest. We waited, as we had been told to, until the imaginary tank came within twenty yards of us, pointed our wooden sticks at the tank, and "fired." Each time I thought about how much heavier a real bazooka would be, and wondered how strong its recoil was. We boys felt we could just as well have played this childish game back in our *Jungvolk* unit. On hindsight, I believe the absurdly deficient nature of the training did not matter that much: military training, even with real weapons, never comes close to preparing one for the situations real battles present.

There was another reason we had been brought into this camp. Our commanders wanted us to join the *Waffen SS*. They believed the uniform of this elite fighting unit would scare the enemy. But I knew that anybody in SS uniform was likely to be shot before they reached the American P.O.W. camps, in response to the rumor—which was not true—that SS units took no prisoners.

I was then just sixteen years old; I honestly don't know how I would have behaved if I had been drafted into an elite unit—whether I would have taken on the unit's attitude and behaved bravely in battle and become a hero; or whether I would have run away to save my life; or whether, if I had been commanded to shoot civilian hostages after my comrades had been murdered by partisans, I would have obeyed orders or refused them. How can one know? War makes all of us different and brings out hidden traits and unpredictable responses.

Certainly my education had left me feeling that fighting bravely mattered. War provided the only opportunity to test oneself in battle. One could fire weapons at the other side and have one's violent acts go unpunished. In war, life was at stake, weapons were to be fired, and one could be killed at any moment.

After the war, I sometimes regretted not having fought in at least one real battle. I felt as if I had climbed up onto the ten meter board and then climbed down again instead of diving into the pool. My war had been spent waiting to find out what a faceless enemy was going to do—whether this time a bomb might find me. It was a suffering without excitement, nothing one could talk about in positive terms.

I would never have joined the *Waffen SS*, though, and in this rag-tag unit of the *Volkssturm*, I just wanted to play dead, but I did not know how. Somebody said the answer was to be sick for a couple of days. He told me he had heard that eating a large quantity of toothpaste would generate a high fever. I was willing to try it.

The next day I fainted during an exercise, and was taken promptly to the infirmary and put to bed. I had eaten two big tubes of toothpaste just three hours before. That was all I could get down. I really felt sick. However, when the nurse made her rounds and gave me a thermometer, I was shocked to see that it showed no fever at all. Fortunately, there was a radiator right beside my bed, so I put the thermometer on it for a moment. When I checked the temperature, it indicated that I would die any second—it read 50 degrees Celsius (115 degrees Fahrenheit). The nurse had almost completed her round: each time I saw her at work elsewhere, I shook the thermometer down. I got it down to 43 degrees Celsius, still a very high fever indeed, and played extremely sick. I was allowed to sleep through at least part of the nightmare that was unfolding outside.

When I was released two days later, the commanders tried to give us a final polish before we marched to our death. We had to drill and practice the Hitler salute one last time, as if approaching the enemy that way would win the battle for us. We had one guy from a small village who seemed to have marched only behind a plow, without ever having lifted his arm to give a "Heil, Hitler" salute. He had to try again and again to march past the sergeant, doing the greeting each time as he passed. He just could not march or get the salute right: it always looked as if he were walking behind his harrow and was waving to a neighbor. We were lying on the ground laughing so hard we developed cramps and would yell, "Please, don't do it again. You're killing us." Even the sergeant finally started laughing; so did the guys whose heads had been shaven for refusing to join the SS. This guy's question-mark posture, together with his strangely outstretched arm, reaching out but sinking below his neck, and his improperly stretched out hand, made him look like a caricature representing the Nazi movement and the state Germany was now in.

[1] Karl May (1842-1912) began writing westerns in German around 1880. His best-known hero, "Old Shatterhand," was an American pioneer of German descent. Winnetou was an Indian,

a noble savage. May's enormously popular books, "mythical-allegorical daydreams of discovery, strife, and redemption," inspired many Europeans to go to America, and have been praised by many who read them when young, including Albert Schweitzer, Albert Einstein, and Hermann Hesse ("the most brilliant representative of a truly original type of fiction—fiction as wish-fulfillment").

2 As more and more adult males were drafted to replace manpower lost in Russia, in 1943 Germany began to enroll teen-age boys as *Flakhelfer* (anti-aircraft auxiliaries). "Often terrified and sometimes killed or wounded, these youngsters 'manned' anti-aircraft guns and searchlights around German cities. Their service contributed substantially to the shooting down of Allied planes; occasionally they found their own homes demolished and their families killed at the end of a night's service. By October, 1944, the boys were being joined by girls." (Weinberg, 756)

3 Starting with Polish prisoners captured in 1939, a great many POWs and foreign workers were brought into Germany during the war, including over one million French prisoners and about a million Soviet prisoners. (Another million Soviet prisoners served as auxiliaries with the German army: more than three million other Soviet prisoners were murdered or allowed to starve.) Another four to five million impressed workers were brought in from the Soviet Union and the rest of occupied Europe, for a total of about seven to eight million forced or enslaved workers. Their labor enabled the government to draft most German men. "Wretchedly housed and fed, constantly harassed and mistreated, brutally punished for real or imaginary offenses, the slave workers were omni-present in wartime Germany. Every town had its slave labor camps, every factory its proportion of slave-laborers, ranging from 20 to 80 percent of the work force ... During the war years, many died of mistreatment, others were killed as 'useless mouths' when unable to work, and the women who constituted more than half the forced laborers from the East were often subjected to forced sterilization and abortions. In the last days of the war thousands were shot on the slightest pretext." (Weinberg, 475-76)

4 Because army officers took part in the unsuccessful attempt to kill Hitler July 20, 1944, the SS was given power over the replacement army, "the new *Volkssturm* or people's army,

the last great mobilization of manpower." (Weinberg, 483) This organization was "a sort of militia for home defense"; its soldiers were given training while still working in factories and fields. It "was to include all males from 16 to 60 who could possibly bear arms if any could be supplied. As the Allies approached and entered German territory from both east and west, *Volkssturm* units were expected to provide local support and reinforcement to the regular army and the armed units of the SS.

7

By March 1945, with Allied armies entering Germany from the east and from the west, it was clear to us that defeat would not long be delayed. Nevertheless, the government continued to function: shortly after my sixteenth birthday, I received a notice saying I was drafted. My mother became very disturbed and desperate; she went to the authorities and told them I could not follow the draft order because I was sick with a heart valve problem. Her plea did no good. This was a time, they said, when everybody was needed, regardless of health or age. (By this point, every able-bodied male between the ages of 15 and 75 was being called to serve).

I reported for duty. We draftees were divided into tank-destroying groups of about eight each. After our farcical training, a non-commissioned officer, a corporal, was assigned to our group, and we were transported by truck to a fighting position not far from Jena. We were moved at night, because the Allied air forces controlled the skies and would attack anything that moved. Even a single individual moving along a road in the daytime had a good chance of being attacked by the twin-fuselage Lightnings the Americans constantly had in the air.

Our group was assigned to block a small road leading to Jena. My first thought was, "But there are many roads like this into Jena, most bigger than this. Why would anybody come through here?" We started digging our foxholes anyway. I had been handed a "stovepipe" bazooka. I had never fired one, but I had learned they were very dangerous, not just to the enemy but to the user as well, because a large flame exited from the back when one was fired. I dug my foxhole in a wooded spot on a slight slope, twenty yards from the road. The ground was very hard; I dug for a day and a half. By then, there was so much loose earth of different colors lying around our foxholes that any oncoming tank would have spotted our position easily, even from a long distance.

We were all very scared. We all knew the war was nearly over, and none of us was in any mood to fight. We knew also that we were in a precarious situation: even if American tanks didn't kill us, a wrong move might still get us shot on the spot. Our corporal or a fervent fellow soldier might kill us for not stepping forward to fight, or someone else who didn't want to fight might kill us while trying to make an escape. Everybody was afraid of everybody else. We watched the corporal closely to see which direction he was leaning. We expected the Americans to show up any day; we could already hear the rumblings of their advance. Nobody stood out among us as a brave, eager Hitler Youth. Everybody was looking for a way to get through the end of the nightmare alive.

Again and again I assessed our situation and tried to envision effective actions: I imagined hearing a tank coming down the road; I would hide deep in my foxhole until the tank was very close and about to pass, and then I would pop up with my bazooka and fire at it. Scenes from childhood playground battles would flash into my mind, and I would briefly get excited about actually living the real thing. No matter how long I practiced in my imagination, though, in the end I always got killed. Either the tanks saw me before I could shoot and fired their guns; or they incinerated me with a flame thrower; or they drove up and buried me alive by turning on top of my foxhole; or, if somehow I escaped all that, I burned myself with the exhaust flames from my own bazooka.

I had just decided not to fire my weapon at all when our corporal called us together. He was sitting with his back against a tree with his rifle on his lap, and was twirling a revolver in his hand, as if he were going to demonstrate to us how to shoot it. All of us sat in front of him, expecting to hear final instructions for the battle that was about to begin. "Listen," he said, "the war will be over in a few days. I am one of the few lucky ones who has made it through this war alive. My experience tells me we all face certain death here if we fire even one shot. I'm going over to that village to see if I can find some civilian clothes, and see whether I can make it back to my family. The war is over, please disperse." We all looked at each other and there was dead silence for a while. We knew it was a situation in which any one of us might be shot instantly, whether he was for or against the order. Fortunately, there were no fanatics among us. Everybody was eager to get away before the inevitable disaster came down the road to find us.

I had never worn a complete uniform, only my black pants and brown shirt. I had a small build and, without a weapon, I looked to be about fourteen years old. I attracted no suspicion walking the fifteen kilometers through the forests back to Jena and had no trouble mixing into the crowd there.

I was afraid to go to my home, though, because everybody there knew I had been drafted. So I decided to rejoin my unit of firemen at Carl Zeiss. In these last days of the war, communications functioned poorly, and, with every available male drafted into the *Volkssturm*, hardly anyone was left to check up on whether people were where they were supposed to be. Only the kids and a few frail-looking older people were left to watch over things. No one at Zeiss ever questioned me about my activities, because I made sure I was always busy doing something.

The very next day after my return Jena was occupied. The day began with a tank battle close to a village named Goschwitz four or five kilometers south of Jena, where there was a big *autobahn* bridge over the Saale River. From our tower at Zeiss we watched the battle with binoculars. We saw one American vehicle drive down a ridge near us but turn back when just one shot was fired. This confirmed our belief that the Americans were saloon soldiers afraid of taking casualties. Then a very slow, low-flying plane passed over, having been sent to see whether any fighting units could be discovered or any gunfire drawn. Not a single shot was fired from our part of Jena. We knew that any resistance would be met with overwhelming firepower, because the Americans had no intention of risking their lives needlessly. This was a totally foreign philosophy to us: Germans were educated to believe that bravery was an attribute of higher worth than even one's life.

Then we observed some American soldiers walking down the road, their rifles positioned to shoot. As they walked past us, we looked into each other's pale and gray faces. We Germans were afraid of the uncertain future, but we were ready for the end of this very long nightmare. Whatever would happen next, we hoped that it could not be as bad as what we had already experienced. We knew that we had at least lived through the war. At that moment I felt as I had after completing a final exam in school: tons of ballast began to lift from my shoulders.

Not long after, tanks and then trucks and many more soldiers passed by. As we looked at their black and white faces, we were amazed by their comfortable-looking uniforms, their almost casual behavior, and the fantastic amount of equipment they brought with them. I loved their drawled American English. I was surprised that soldiers talked like that to each other, in tones and cadences so different from the staccato German military communications I was used to.

The American soldiers took over some houses, ordering the residents to move out. They lived in them pretty much, I know today, as Americans live at home. They gave us kids, especially the younger ones, tons of chocolate candy and chewing gum. Since I did not speak a single word of English, I could not communicate much with them. I remember asking a tank crew what type of tank they had, but they did not understand. I kept repeating, "Is this a Sherman tank?" but since I pronounced it almost like "German" tank, they always answered: "No, buddy, this is no German tank." I also could not pronounce *chewing gum* correctly, saying something more like "jewing gum," which never got me any.

One day I was sitting in front of an American bakery watching the GIs unpack and carry in a truckload of American flour. One soldier, who must have seen my hungry look, threw a twenty-five-pound sack of flour over to me. I wish to this day I could meet this soldier again and thank him, because we lived for almost three months off this bag, cooking a white flour soup every morning for breakfast. With a little salt, it tasted terrific.

For some weeks after the occupation began, we had strict curfews at night. Then life slowly returned to normal. People began cleaning up the rubble of war. Older men with disoriented faces, fragile from deprivation, shoveled the debris off the streets. Women, pale, worried, and worn-looking, dug out bricks left in the ruins of houses and cleaned them for reuse. All of us were grateful for having been freed by the Americans, though we didn't like having their soldiers wanting to sleep with our women. I remember people got especially upset when some woman brought home a black soldier: after all, we had been taught that they were racially inferior. Almost the only blacks we had ever seen had been in movies set in the former German colonies in Africa, and those people had been depicted as ignorant natives. We were so ignorant ourselves that we thought their black coloring might come off on the bedsheets.

In the Third Reich, German women had been raised to semi-holy status as the ones destined to give birth to a new, even more Germanic-looking master race. After the birth of her fourth child, a woman could even receive the *Mutterkreuz* (Mother Cross), having made her contribution to the mission of producing enough soldiers to occupy the entire world. If she continued to serve her country through her fecundity, she could even achieve the golden *Mutterkreuz* for her twelfth child. German women sleeping with occupying soldiers provided a very elemental representation of the total defeat Nazi Germany had suffered.

Overall, however, we liked the Americans and got along very well with them. More surprising, we even started to adopt some aspects of the American culture, a culture which we had been taught to despise. You saw quite a few people imitating American informality of dress, or attitude.

Six weeks after Jena's capture, rumors started to surface that the Americans were going to leave and that the Russians would be coming to occupy us. We found this unbelievable. How could the Americans give up territory they had spilled their own blood to conquer? As we found out later, the boundaries of the occupation zones had been agreed to at the Yalta Conference. The Russians got Thuringia, where we lived, and nearby Saxony; the western allies got zones in Berlin.

Soon the Americans did leave, taking with them from Carl Zeiss and Schott sixty scientists and specialists, and all the special-purpose machines, without which certain products could not be built. Now the full color movie we had been briefly allowed to live in turned into a dismal black-and-white one which would not end for more than thirty years. We sensed how dark it might get when we saw the first Russian troops arrive. They were from another world. They all looked exactly like the very poor Russian P.O.W.s we had seen in Jena during the war. I don't remember seeing a single motorized vehicle. They came on foot, their belongings and weapons piled on little wagons drawn by horses. We watched in disbelief. Could these be the people who had beaten the German army? Seeing these soldiers, we felt as if history had been turned back hundreds of years and we had been shifted a thousand miles to the east—to the places the German Wehrmacht had wanted to conquer for us just a few years ago.

The whole population was scared, almost in panic. There would be no escaping this. One could only hold on to oneself and hope to retain one's belongings. After what these soldiers had gone through, they needed everything, and they would satisfy their hungers by devouring everything we had. No scruples restrained them from raping our women—which they did. For them, after their experiences, raping a woman was nothing; even the thought of a woman resisting seemed ridiculous.

Even though the Russian army had defeated the German army head to head, we still thought of them as racially inferior. Hitler had called them *untermenschen* (subhumans). We hated their *kultur,* which they wanted to impose on us. They installed loudspeaker systems on corners throughout the city, and we had to listen to them playing their songs and yelling instructions over them. The only outlet we had for our hatred came in spreading stories about anything bad or ridiculous any of them did. We eagerly passed along the report that a Russian soldier exchanged his expensive watch for a cheap one because he didn't know he had to wind it. We repeated the tale that a Russian soldier who saw a German riding an old bicycle without holding the handlebars insisted on trading his new one for it because his only worked if he held the handlebars. We reported that Russian soldiers washed their potatoes in our toilets, and in fact some of them may have. Certainly a lot of the Russian soldiers came from rural areas where they might never have seen indoor plumbing. We never saw these soldiers as people, never thought about the villages they had come from, or the schooling they had had, or the families who were worrying about them.

We had to feed the Russians many of our possessions—watches, bicycles, radios, and even cloth. We had to feed them houses, and then more houses: they needed new ones every so often when the old ones grew dirty and sometimes full of excrement. Worst of all, they kept eating people: they would just seize citizens right off the street and deport them to places where they were needed, and from which they would never return. The Russians required every German to be thoroughly "de-Nazified." They questioned a citizen; if his activities during the Nazi time were unacceptable, he would take a trip to the Soviet Union to P.O.W. camps where he would be "de-Nazified" again and again until he died. Being sent to work in the radium mines had the same effect.

All administrative positions were occupied by either Russian or German Communists. Many *Kameraden der Roten Front* (Comrades of the Red Front) had spent the war in Nazi concentration camps or in Russia itself. We did not believe for a minute that these inexperienced people could manage the country or its economy successfully. They were lower-class people with very little education; we were used to having our country be run by intellectuals, or at least by well-educated people.

The food supply was now even more meager and irregular than it had been during the war. Clothing and footwear no longer could be found in any store. Yet somehow life went on. People went back to work at the Zeiss and Schott plants. I continued my apprenticeship. I even went to some of the meetings of the new "Free German Youth" organization. The new authorities generously made the big villa of a former Nazi available for our meetings. It was almost like the *Jungvolk,* only with the political polarity reversed. German-Soviet friendship was trumpeted everywhere. Suddenly there were more Communists and fewer Nazis than the most cynical prophet could have foretold. I think that for a short while I even became infected by communistic ideas. Sharing everything (I had nothing) with everybody else made sense to me. But when I thought things through, and noticed that everyone was telling the Soviets the lies they wanted to hear, my sympathy for collectivism ended.

A great many strange transformations took place in these months. Many Germans rewrote their own past. The Russians rewrote their history for our consumption to assist us in believing they were a great, cultured, and well-meaning nation, one which had invented every piece of modern technology well before the Industrial Revolution took place in the West. Russia, they told us, was not a nation of farmers: technology had penetrated even the cow barn. Their propaganda films showed people wearing white coats in a dairy so clean one could eat from the floor among the cows.

As for us, we continued working sixty hours a week, exactly as we had during the war. The difference was we did not work as hard, because we often had nothing to work on, and also because we knew all our output was being shipped as reparations to the Soviet Union.

The Russians brought a great many engineers and scientists into our plant to study processes and tap into our know-how. We smugly believed they could never in a million years learn all the things time had taught us. We felt very safe.

At work we were constantly bombarded with Communist propaganda about how good everything was under Bolshevism and how good everything would become for us if we just bit the bullet and worked hard. But nobody could; we had almost nothing to eat to sustain what little strength we had, and there was nothing we could buy with the money we earned.

I was so starved that I would walk into nearby villages and ask the farmers for food, but they had hardly anything themselves and lived under the constant threat of losing their farms. Eventually many of them did lose everything they owned. All the larger farms were taken over and broken into small farms, many of which were given to people who had never been farmers and had no idea how to grow crops or raise livestock. All this brought the Eastern zone, the territories occupied by the Soviets, close to starvation. During the worst wartime air raids, we could at least expect to hear at some moment a siren that would tell us the attack was finished, but this nightmare seemed as if it would never end.

We also had almost no fuel. No firewood or coal could be purchased in the stores. The forests had been swept clean; every branch had been picked up. People even had to keep watch on their garden fences, because if the weather became really cold, foragers would just rip out the boards and take them home. That year, 1946-1947, we had a very cold winter. After suffering for weeks and weeks, I made up my mind to do something drastic. I took off in the middle of the night, saw and ax in hand, walked up the road to the forest, and began cutting down a tree that I thought was small. Without experience, however, one does not know how big a tree is until it is down. This tree turned out to be big and extremely heavy from the spring sap already in it. I worked until early in the morning pulling the tree down the snow-covered paths and streets to our place. It was an extremely dangerous operation, because at that time you could get shot if you were caught taking anything.

The food situation kept getting worse. Often we had nothing to eat for a whole day or two. I was seventeen then and could eat like a horse if I had food. Every weekend I walked out to the farms begging for a job to earn a meal, or some potatoes, or even better, some wheat or corn or cabbage to take home to my family. Eventually I managed to get a bicycle from an acquaintance. It was very old and shaky, but it still had some tread on the tires and was good enough to get me up and down the country lanes. A few times I rode out through the forest to a corner near the fields, carried my bicycle into the winter wheat fields so as to leave no tracks, spread out a blanket, and cut wheat into small bundles. I then positioned the rear wheel over the blanket and, turning the pedals with my hand, threshed out the grain by holding the bundles in the moving spokes of the wheel. Using this method, I was able to thresh out about half a bucket of wheat that I later separated from the chaff at home by throwing everything across the room and sweeping up the kernels. We then put the grain through a small hand-operated coffee grinder and ended up with a coarse flour we used to prepare flour soup. It tasted very good and kept us going. Another time, when I was about to come home empty-handed from the country, I stepped off the road into a potato field and dug out some seed potatoes that had already sprouted plants a foot high. When we tried to cook them, though, we found that we could not get them to soften no matter how long we boiled them.

When I traveled to these farm villages, I always carried a short, thick stick with me as a defense against the vicious dogs the farmers kept to sic on guys like me. On one trip, while riding my bicycle near a village, I spotted a chicken on the road ahead. I threw my stick at it from very far away and hit it on its breast. Stunned, the chicken was unable to make any sound, but still had the strength to run off the road and disappear through a gap under a fence. I thought I had lost it and was about to go on when the chicken suddenly came running back out, and I was able to grab it. This was the highlight of all my food-hunting trips.

Looking for food became an obsession, occupying almost every spare moment. Hunger drove us to pursue any avenue. One friend had seven younger brothers; all of them were starving. Once he stole a potato from a pot of boiling potatoes and popped it into his mouth to get it out of sight quickly. He burned his mouth very badly. Like an American trapper in the northern territories, he set traps for rabbits in a wide circle around his neighborhood. Every morning before dawn he made his rounds, checking his snares. And he caught quite a few rabbits. He would have eaten any other animal he caught, as well.

Around this time I received a very big lesson in "crime and punishment." In another part of the town I went into a house that had a bakery in its ground floor. I picked up two loaves of bread and walked along the long aisle toward the front door, holding the bread in front of me. A baker noticed me and began yelling as he ran after me. He chased me more than a mile. I ran, crazy with fear, holding the bread in my arms. I thought I could outrun him, but the bread slowed me down, and I could hear him gaining on me. Eventually I heard the baker right behind me, so I spun around, threw the bread in his face, and began running again. He followed, but without the bread I was able to put some distance between us. I dodged around a few corners and into a large building. I took off my shoes and ran up the stairs all the way to the top. In the area just under the roof I hid under a big woven wicker laundry basket. I hid there for two hours, praying and swearing to God that I would never commit another crime if He would allow me to escape punishment for this one. He listened; I eventually walked away unnoticed.

When my cousin came over one day, starving, we agreed to kill the next cat or dog we could catch—but there were no cats or dogs around anymore. Few people could afford to feed an animal at that time, and before we woke up to the possibility, other people had already had the idea of butchering and eating them.

Six weeks went by after my cousin's visit and I forgot about our plan. My grandmother went away to visit her mother in Kaltennordheim. Then my cousin showed up with a sack. He had caught a cat in it. I was stunned, apprehensive, and scared. But the Third Reich had imposed on me a code that meant, above all, never going back on one's word. I had no choice but to go ahead and kill that cat. I had never done such a thing before and did not know how to proceed. I also loved cats. I finally pulled myself together, took a big wooden meat hammer from among my grandmother's kitchen utensils, and gave the cat a wicked blow on its head. The cat made the most incredible jump I had ever seen a cat make. It jumped all the way from the sofa to the top of the tall kitchen cabinet and then kept jumping everywhere, crisscrossing the room until it had exhausted all its nine lives and came to rest on the floor, dead. My grandmother's kitchen was a mess; there was blood everywhere. We skinned the cat, cut it up, and gave most of it to my uncle and aunt, saying it was a wild rabbit we had gotten from a man who trapped them. I had no idea how to cook my portion. I decided to fry it in a pan. Hungry though I was, I had great difficulty eating even a few bites. When my grandmother learned what I had done, she did not speak to me for weeks, and she never used that frying pan again. I spent a long time thinking about that cat and mentally put myself on trial whenever I saw another one.

By 1947 all my relatives had returned home. My cousin Karl had come back long before the end of the war. Badly wounded, he had been evacuated from Stalingrad on one of the last planes out. My brother Frieder had been near home by the end: he had surprised us by deserting his paratrooper unit in 1945, right after he was handed a brownish uniform that looked like the SS uniform. He told me, "When I saw that uniform, I knew I had to escape, because if I got captured by the Allies wearing that, I knew I would be shot." He had hidden in the forest near Jena, and we had brought him food. Uncle Alfred had been a P.O.W. with the Americans. My Uncle Helmut had walked back from the Balkans. He had made it back alive only because when his front kino unit had shown movies to the troops, he had let a lot of the local people watch, too. The goodwill he created paid off for him: the partisans there had let him go free, while they shot many of his fellow soldiers.

My mother's husband was the last one to come home. When he rang the doorbell and my mother looked down, he said, "Paul is back!" I remember my mother turning to me with a frightened and disappointed look on her face, saying to me, "He is back." Their marriage had never worked well. They got divorced two years later.

The "Free German Youth" meetings had been worth something: I had learned to play the guitar. I practiced often, and I had a pretty good singing voice. People began to invite me to sing at weddings and special occasions. I was interested in girls, and this was a good way to meet them, but it was a long time before I kissed one. She was a girl from the neighborhood who was very smart and had lips I could not resist. Her name was Hilde Messer. She translated her name into English for me and it became one of the first words I learned in that language. It meant "knife." Our relationship remained innocently affectionate for a long time.

I was still very much focused on becoming a good journeyman in optics and mechanics. We apprentices had three hours of professional school every day, and in addition to developing my technical skills, the schooling helped me improve my writing. I had never paid much attention to German spelling and grammar before; in fact I had been so bad in German in the *Volksschule* that I almost did not get promoted. But now we had to write each day about what we had learned. Between that writing and the many books from libraries that I read during this period, I managed to catch up in my facility with the German language.

Suddenly one night a big shock came: a whole army of Russians drove into Jena, bringing many big trucks. They banged on doors and rounded up 600 specialists, managers, and scientists, together with their families. These people were allowed to take some belongings, even some furniture, with them to the railway station, where a train was waiting. They were herded onto and locked into that train, and put under heavy guard. Two days later they were taken off to the Soviet Union. We learned that they had been assigned to build duplicate plants—factories exactly like those of Carl Zeiss and Schott—right outside Moscow and Leningrad. We also learned that our entire plant, everything, was going to be dismantled and shipped east, too. We could not believe this cruel development. What about us, we asked each other? What were we going to do without a work place?

It began to dawn on us that we had been outsmarted. The Russians were, after all, not as dumb as we had thought. They had allowed us to maintain our smug, superior, arrogant attitude and had turned it against us. Their engineers had infiltrated the plant completely and had studied everybody. They knew exactly who was needed in Russia to rebuild Zeiss.

We had no choice. We crated up everything for the move, but it seemed to us an absurd undertaking. How would they get all that equipment, especially the very heavy machinery, on the trains? We did not believe an entire factory as big as Zeiss could be shipped to Russia. Again, the Russians showed us they were smarter and more clever than we thought. They had us mount the machines on wooden beams that had been sawed off at the end at an angle. They then pulled these like sleds down the road to the railway station. Sometimes the wood got so hot from the friction it would start to burn. But the fire would not spread because the beams had been cut from green and juicy trees just days before. The 15,000 of us who were employed at Zeiss spent nine months dismantling our factory. The Russians were taking everything. Even the old parquet floors were ripped out and packed up. The toilets, including the uncleaned waste pipes down to the collection tank, were wrapped in oil paper and shipped to Russia.

But they were not able to take everything. We hid many thousands of tool bits, gauges, cutters, and other tools by throwing them into a junk pile in the factory yard. Some employees pushed special machinery down into the many underground tunnels and walled them in with bricks and concrete. We planned someday to rebuild our factory. We felt it was *our* factory, because the Carl Zeiss plant, about one hundred years old, had been given by its owner, Herr Zeiss, to its employees as a trust. Each year some of the company's profits were put into a fund; its size depended on how well the company had done that year. Every Christmas each Zeiss employee received a share of that fund proportionate to his wages. Carl Zeiss's concept had been far ahead of its time and was unique in Europe. The Russians ended this system. There would be no "worker's capitalism" in East Germany; everybody was to be made equal.

We workers did not feel equal now. We had no work place, no future, and nothing to wear but rags. I did not even have socks or leather shoes. I walked the long distance to the north train station every day in shoes with leather tops riveted to wooden soles. Before I put them on, I artfully wrapped rags around my feet and ankles to protect them from getting blisters and from the biting winter cold. Every so often, I went to the open fire we maintained on the street to warm up. We had modified oil barrels to become heat-stoves, in which we burned everything we could find. Sometimes we even burned the oily parquet pieces we were supposed to ship to Russia. In all, it was one of the hardest and most degrading times I have ever experienced, but I discovered I could endure most all of it rather well, even the cold and hunger. Sometimes, to escape the bleakness of the situation, I would imagine I was on an expedition to the North Pole. But I was depressed by our total helplessness in the face of another power, a superior state that kept telling us it was our friend as it robbed us of the necessities of life.

I reached bottom one of the nights when several of us had jumped onto a moving train in order to throw coal off the cars. We collected this coal later from the railroad tracks. On that night my girlfriend Hilde was caught by the Russians and raped by five of them. This was how she lost her innocence. There I stood in my wooden shoes as if I had been raped myself. And I had been—all of us had been.

8

In 1946 the Soviet military government had given some control of politics in their occupation zone to German communists. This new government had decided to rebuild our factory, and we were making good progress. I did all kinds of odd jobs to help: one was to dig through the junk piles to recover the tools we had thrown into it while the Russians were dismantling everything. Most of the rest of my time was spent making all kinds of metal parts for tooling and instrumentation. I had finished my apprenticeship with very good grades: I had come a long way from Loebderstrasse. Proud of having become a skilled worker, I thought then that I could earn enough money and perhaps live a good life and be happy forever right there at Zeiss.

But one could hardly buy anything with money—not even food was available. For a while I had an arrangement with a baker's daughter who was much older than I. I manufactured finger rings for her from some old stainless steel cores left from core-drilling the periscopes for U-boats. They were just the right size: I would cut, turn, and mill the shiny metal, and then file a dovetail notch into the top to receive a stone. I made the stones out of colored chips of glass, cutting, grinding, and polishing them to have facets that sparkled and a base that fit snugly into the notch. It took me three weekends to make a ring, which I then traded for several pounds of bread. I suspected that she stole it from her father's bakery.

But my little market quickly became saturated with product, and I returned to my former method of adding to our stores—my trips into the countryside. They became almost a business. I bartered a lot. I would get some fabric, my mother would sew aprons out of it, and I would trade these for food. I would also file new keys for the farmers while they waited or sell items left by people's deceased relatives. I became very good at this trading and brokering, and so we always had at least something to eat.

Once I fabricated some big flat springs for some farm equipment. This undertaking was incredibly difficult because I had to do everything secretly at Carl Zeiss: obtain the half-inch-thick metal sheets, cut foot-long, two-inch wide strips from them, finish the sides, drill holes, and bend the strips into a complex shape. I was able to trade them for a lot of food. I did not go back to that farm afterwards because I knew the steel I had used was not suitable for springs: it could not be hardened and I had to leave it in its original soft state. Much later, I happened to encounter that farmer and he started waving to me from afar. I thought to myself, "Now you are going to get it. He is going to kill you." But to my surprise, what he said was, "Where have you been? I need a lot more of those springs you sold me."

I said, "Really?"

"Yes," he said, "those are by far the best springs I've ever used."

I said, "Yes, but they are soft."

"Yes, and that is why I like them," he answered. "When I hit a rock with my equipment with ordinary springs, they break, but with yours they only bend and I can just bend them back." This surprise testimonial showed me how hard it can sometimes be to recognize that what you've created is a winner.

Business got better once I rode far enough from Jena that the farmers were not being constantly pestered by begging townspeople. After a year or so, though, I began to worry that I would have to end my forays because my bicycle's tires were about worn out. I couldn't get new ones, so I worked to extend their usable life. I tried to protect the inner tube from getting punctured by putting pieces of old tires over the holes in the tire and then wrapping flower wire around the tire and the rim. Every fifteen miles or so, the wire wore out and I would have to stop and replace it.

On one trip to the country I met Rudi Berlin, a giant of a man who was doing the same thing I was doing. Rudi told me that he had obtained drawings for an oil press which could extract oil from all kinds of seeds and that he was now busy making such a press. He had also just acquired an old motorcycle, an Ardie 500 with a sidecar. It had been made in the early thirties and looked like a big elk. When we got ready to repair it, we could find no place to do the work other than his apartment on the fifth floor of an apartment house. We became the only persons ever to drive a 500cc motorcycle from the street up to a fifth floor, and then back down to the street again. We put a thick plank onto the staircase, and, while I held the back in balance, Rudi drove the machine up one flight at a time with earthshaking noise and thick smoke. This was harder than it sounds, and very dangerous: it took us a whole day to get the motorcycle up, and another day to get it back down. We were very lucky to have this motorcycle. At that time one rarely saw any cars or motorcycles in or near Jena. People even rode bicycles on the *autobahn*.

I worked day and night to help Rudi fabricate his oil press, and became his buddy. The job took a long time, but after the press was finished we made many trips with his motorcycle to the surrounding villages, and pressed out a lot of oil, which was very hard to get legally. The deal was ten bottles of oil for the farmer and one for us. Private trading was illegal, with severe penalties, but we and the farmers both needed oil so badly that we were willing to take the risk.

The oil opened new possibilities for bartering: for one or two bottles of oil we could obtain a pair of very good used shoes or any piece of cloth we wanted. It was now 1948; I became, for that time and place, a rather well-dressed young man, advancing myself a step further and gaining more confidence.

I was now almost nineteen, and I had never traveled far from Jena, except for the train ride to my great-grandmother's village years before. I wanted to see what the rest of the world looked like—and seeing more of the dismal Russian zone was not what I had in mind. I wanted to explore the Western zones and see big mountains—the Alps, if possible. I did not think then about leaving East Germany permanently, for I felt deeply rooted; all my relatives and friends were there. I did think, though, of how my father had walked the lands in his Jesus shirt, and I was willing to sacrifice some time and money to do something similar. I did not know then that this first journey would begin a lifelong pattern for me.

It took me a long time to collect what I thought I needed for such an undertaking. The final missing item was a pair of shoes that could withstand wear and weather. Then came a day when my work place gave me an allotment slip for one pair of shoes, and I found a good, tough pair of work boots. I was now ready, but I still hesitated. I knew no one else who had even thought of doing such a thing; most people were too poor, too cautious, or too scared. It helped that my girlfriend Hilde and her parents had packed up their belongings and moved to the West. I knew I could stay with her for at least a while. Two or three months later I confided to my boss, Herr Sittner, that I wanted to quit my job and go exploring. He told me, making sure nobody was listening in, that just between us he considered it a good thing for a guy my age to do. "You will learn a lot," he said. "Just make sure you don't stay with your girlfriend longer than two or three weeks; otherwise you might get married, and you are too young for that."

So I quit my job, put on my backpack, and, with my guitar hanging over my shoulder, took off for the West. I rode a train almost to the border, waited until night, and started marching south, orienting myself by the moon. The border of the Russian zone was, at that time, the only border of the four occupation zones that was completely closed and illegal to cross. There was no fence, but Russian soldiers guarded it on our side. At that time people crossed secretly from East to West in rather large numbers. I didn't really know when I had crossed the border, but when dawn came, I ran into an American border patrol that picked me up and brought me to its shelter for interrogation.

The American soldiers asked where I wanted to go.

I replied: "To my girlfriend in Zwiesel, in Bavaria."

"What are you going to do with that guitar?"

"I want to serenade her," I replied.

And so it went as they examined everything I had with me. They even made me sing some songs for them. They were not convinced I was telling the truth until they reached deep down into one pocket of my backpack and found a few condoms. They found this very funny, and we ended up all laughing together. They even offered to drive me to Bamberg to the railroad station. This is how I was received in the American zone.

I caught a train and arrived late that evening in Zwiesel, where Hilde lived with her stepfather. The next morning I had my first opportunity to assess life in the

Western zones. Evidently many people here were just as bad off as we were in the East. Hilde lived with her parents in little barracks-like houses without real inside walls. Rooms were separated by partitions made of the same type of coarse brown paper with an oil coating on one side that we had used to wrap up everything for the Russians. I had brought a bottle of oil along, and we used it to fry potatoes every day until it was finished. Hilde's stepfather did not really welcome me, however; I thought he was after Hilde himself.

For all these reasons, but especially remembering Herr Sittner's advice, I said goodbye to Hilde after one week and left for destinations unknown. I visited all kinds of towns and villages across Germany, my itinerary dependent only on which train I could catch to any place that sounded interesting. At that time, most trains were packed, and one had to force one's way in. People often hung onto the outside of the train or stood squeezed together on the open platforms between cars. No ticket collector could get through to ask for money; and if he had, some desperate individual might well have thrown him off the train. These were tough times—nobody had much to lose, and a lot of people rode without tickets.

I carried 3,000 Reichsmarks in a bag tied with a string around my neck and hidden under my shirt, but I never had to touch that money—there was nothing to buy with it anyway. My routine was to go to a town, walk through it and look at everything, and then go out to a village to look through the help-wanted notices on the notice board in front of the church. I always found some work to do and so was able to keep myself fed. I became a jack-of-all-trades, learning how to do pretty much everything that I had not learned earlier through my apprenticeship. I learned most abosut farming—how to take care of animals, to milk, to make hay, to sow wheat, to repair farm machinery, to handle horses, and to walk behind horses, tilling the earth with a plow. I felt great; I loved this work out in nature. I met a lot of people from other walks of life and learned to understand different dialects that were as strange to me as any foreign language. Sometimes I wondered how much of what I was experiencing my father had seen and sensed years before.

Occasionally I did not make it out of a city by nightfall and had to sleep in a park or even in shelters for the homeless. (There were not many youth hostels at that time.) I always took precautions against having anything stolen. While I slept, I kept everything fastened to my limbs. Since my shoes were my most valuable possession, I always tied them together under my head. I loved talking to the people I met in these places, most of whom were very nice. Some were just between jobs, like me, while others chose to live their lives as bums or tramps. When I later worked on semi-automatic machines doing piece work, living most of my day chained to a treadmill, I often thought about these people—how relaxed they were, how they might even then be going to the park or listening to the stories of newcomers. I never met one who felt burdened by obligation or responsibility. They had done away with all that.

I wanted to see big mountains, and so one day I caught a train going south toward the Alps. I customarily traveled standing near a window to observe the landscapes and study the crowds at the railroad stops. When I arrived at Oberstdorf/Allgäu and saw all the mountains there with snow on them, I felt like the happiest person on earth. I decided immediately to stay a long time in this wondrous, fantasy-like world, find some work and start climbing mountains as soon as I could.

I went as usual to the church notice board to read the job notices. One little handwritten piece of paper read:

"Looking for farm hand to work in Austria." I was elated. There were even higher mountains there, and it was a foreign country even further away from home. I had to get this job. I almost ran to the address indicated. It was a beautiful, rich-looking house. An attractive woman in the colorful local costume opened the door and greeted me in a friendly manner. Pulling all my charm together, I asked for the job. She explained to me that her brother in the Walserthal needed a strong hand and would be coming down in a few days. Meanwhile I could help her in her garden. Thus I started to work as a gardener, with full room and board in this wonderful house. I was ecstatic. Nobody going on a great expedition, even to Africa, could have felt more excited.

After one week, Pepperl Kessler, the man who needed help, arrived in his Volkswagen to interview me. I liked him immediately. He looked as tough as hammered steel—a man one could count on when climbing in the mountains or in trouble. I could also see that he was as unspoiled as nature and somehow I knew he would play a major part as a role model in my life. This man did not look like one of the regular farmers. His eyes had fire and goodness in them, while his posture and agility showed that he was a great sportsman. He looked tough and tenacious, and seemed to have been around many types of people other than the locals I had seen in the village. In fact, Pepperl had been an officer with the mountain paratroopers and had served during the war in the French Alps. He was also a certified mountain guide and one of the great ski champions and ski instructors of his valley. Fifteen years older than I, he had the strength of a bull and could, as I soon learned, lift a 500-pound stack of hay from the ground onto his shoulders and carry it over steep terrain into the barns. He also had completed an apprenticeship, was a terrific mechanic, and was generally a man who could do anything. It was curious to me that such a powerful, talented man had, nevertheless, a significant problem to contend with: he stuttered badly.

Pepperl told me he had a very tough job for me. His farmhouse was high up and isolated in the mountains, and the meadows sloped steeply. Everything there had to be done by hand without the aid of machinery. I said only, "Yes, yes, yes." I wanted to go and see that place. I remembered an important comment one of my teachers had made to me when I graduated from my apprenticeship with Zeiss: "When you apply for a job, always reach a little higher up than you think you can perform. And if you get the job, work like hell to live up to what you have promised. Put everything you have into it and perform."

When I went with Pepperl, I had no idea how far I would have to stretch myself to live up to this man's expectations. First, though, I would have to shrink. On the way to his place, Pepperl told me there was no way to get me into Austria legally, so he was going to smuggle me into his country. Before we reached the border, Pepperl made me lie down in the back of the car and started piling all kinds of things over me. I heard him stop at the border and talk to the guards, who must have known him. They let him proceed on into the valley. He lived in the small village of Hirschegg, which consisted then of a church and a few farm houses scattered over the grassy hills. Above the meadows were forests, then meadows again, and then bare rock. It was the most beautiful place I had ever seen, and Pepperl's farm far up on the hill had the best view of all.

The farm actually belonged to Pepperl's parents. Pepperl and his whole family worked the farm. His parents looked at me skeptically. They judged me to be a small, weak foreigner who had never done farmwork and could not possibly do the job. Comparing myself to Pepperl, I could understand why they thought this.

I wasn't paid much—fifty Deutsche Marks a month. That didn't matter, because money wasn't of much use then. Some things couldn't be bought because no one had them to sell. I myself couldn't even buy the clothing or food that was available: since I was in Austria illegally, I could not get the rationing stamps one needed at that time to buy almost anything. Like everyone else, I made do with what there was. For example, by the time I reached Pepperl, I had worn my good leather shoes all to pieces, and I couldn't buy replacements. So Pepperl gave me some old shoes of his to use when we had to work in the meadows, which were very wet. The shoes didn't fit, but I wore them anyway: I had no alternative.

My workday started at 4:30 in the morning. I would get up fifteen minutes before, go to the front of the house, wash myself in the ice-cold water coming right off the mountain, walk to the barns, and shovel the dung into a precisely stacked pile. If the job wasn't done just right, the dung would have fallen down the hill. Then I had to spread new straw under the cows and feed them hay. There were about thirty cows. Pepperl never let me milk them because he was afraid I would not milk them completely and they would consequently give less milk. Instead, I had to carry the heavy milk cans, three at a time, one on my back and one on each hand, up the hill to the house. They weighed tons, and for that job I wore wooden shoes, similar to the ones the Dutch make. They slipped, especially when it was raining. We poured the milk into dishes to let the cream rise to the top, and then hand-processed it into butter or cheese.

Then it was time for breakfast. We usually ate a big pan of fried potatoes or homemade buttery noodles. Everything was prepared over an open fire. We all ate out of the same pan, everybody eating a wedge toward the middle. I always got to the middle first, and the others, seeing I was still hungry, would often leave me a big portion of their wedges.

After breakfast, the actual workday started. We would build fences, or fertilize all the meadows by taking the cow dung there and hacking it into small pieces, or cut down very big trees in the forest far up above the house and bring them to the house. Or we would cut the grass and turn it out to dry, or bring some cows to the bull, or repair the house and barns. We stopped only to take our meals, and continued working until ten or eleven at night. Late at night we usually made ax shafts or shingles for the house or did other woodwork.

I had never worked so hard in my life and often thought I could not sustain it, but I managed to do what Pepperl expected. I also never ate so much in my life. His parents continually worried that they were not getting a good enough deal with me as their hired hand—they kept saying they had never seen anybody eat like that. And in reality they did not see everything I did eat. Secretly I would drink gallons of milk before I poured it into the cans. I would sneak into the cellar and eat the fantastic homemade Swiss cheese by the pound, and some apples, too. As a result, I gained fifty pounds in ten weeks and ended up looking like a miniature Arnold Schwarzenegger, all muscles and deeply tanned from the mountain sun. We worked seven days a week, but on Sundays Pepperl and I took time off to go mountain climbing. I always loved these little trips.

After I had worked there awhile and had learned how to do things correctly, I was given an assistant. Our relationship was unusual: my "assistant" was the German General von Wiedersheim. He was lying low in Austria, waiting to be called for interrogation about his actions during the war. Until he had been questioned and had been officially certified as de-Nazified he could not get a job that used his education and skills. Like everybody else, he was in need of food, and he was willing to work for it. We did many jobs together. When he arrived each morning, I would greet him with, *"Guten Morgen, Herr General,"* bending down my upper body, as if he were a king, or standing at attention, clicking my wooden shoes and lifting my hand in a military salute as if he were still on active duty. Then I would give him the day's instructions. Even though we both wore ragged clothes and stank worse than the cows, he still behaved as if he were attending a social event at an officer's club, and I developed some class in trying to emulate him. It was a great privilege for me to work with this very astute man.

There was, however, one situation in which I felt his behavior fell short of honorable. I had been given a spade and told to till a little garden overgrown with large weeds. Instead of pulling the weeds out, I had buried them carefully, shovelful by shovelful, under the surface. When the general was assigned to seed and plant, and started to work the soil with a rake, he quickly found all those weeds and had to pull them out. He told the older folks about it, and they got quite upset. I heard Pepperl's mother yelling, "He is more stupid than a cow!"—meaning me, of course.

They would have said the same thing had they known about the rooster episode. The farm had the biggest, most aggressive rooster I have ever seen. Every visitor and even the mailman was afraid of him. He would attack anyone, flying up from behind you to attack your neck and head with his sharp spurs and beak. Having this rooster around was downright scary—the only way to avoid being attacked was to keep an eye on him at all times. After he had gotten me once too often, I decided to teach him a lesson. I grabbed him by his neck and dunked him for several minutes deep in the cesspool. When I brought him back up he looked terrible, his feathers wet and full of excrement. He appeared to be dead. I took him to a place where nobody other than a fox could find him. Then I forgot about him ... until he jumped me again the next morning.

All the local farmers used toilet wastes to fertilize their pastures. Most of them carried buckets up to the meadows, or had their ox pull a wagon. Pepperl had devised a way to use a winch system to pull a wide two-wheeled cart with an open container in it. No one else in the valley had mechanized this difficult, time-consuming, and laborious task. From the winch near the farmhouse, a cable ran up the mountain to a pulley, and back again to the cart. We moved the pulley when we wanted to fertilize another area of meadow. While one of us guided the cart by holding its handles, the other would sit on the winch motor, turning it on and off and controlling the clutch as required to pull the cart up the hill. One day I was guiding the very heavy cart up the steep slope. Everything was going well, and when I became exhausted, Pepperl put me on the winch. The cart was fully loaded. He yelled over to me, "Go!", and I turned on the motor and engaged the clutch. He started to walk behind the cart, holding the handles and guiding it up the very steep meadow. As he passed me, he looked over and said, "Don't let go!" But his dialect changed the word order to "Let go not." This would have been fine had he not been a stutterer; but he made a long, painful pause between "let go" and "not,"and before he uttered "not," I let go of the clutch. The cart tilted instantly downhill, sinking the handles into the soft ground and shooting a whole waterfall of sewage over Pepperl. It was one of the most hilarious scenes I have ever seen. We could read newspaper articles from many months back on different parts of his body.

After several months, all my clothes were worn out and my shoes would no longer hold together. My mother had written to me as well, saying she was ill and needed me. I came to realize that I had climbed enough mountains for now, and that in my time with Pepperl I had learned a great deal, especially how to work incredibly hard, and had acquired some insights into how the world outside Jena worked. As when I trained with the wooden bazooka, I had prepared myself for the battle to come—but in this instance, far more effectively. I was ready to stage another sort of climb to reach the higher regions where Pepperl and the general already dwelt.

I packed up my few things, and Pepperl drove me close to the border. We had become good friends. We said goodbye affectionately, and I promised to return soon, to ease Pepperl's pain at losing his helper and *kamerad.*

I walked through the forest, bypassing the border guards. It was a beautiful, clear, starlit night. Everything went well, except that close to the guard station I climbed over a cow, thinking it was a rock. The cow and a few of its companions jumped up and made a racket with their cow bells. But I was already on German soil, on my way home to Jena.

9

Early in the morning I rang the doorbell at home. My mother did not recognize me at first. I had returned grown into a man with broader shoulders and with an incredibly dark tan. She was very happy to see that I had survived, but she had to leave right away for work. I had not eaten much on my trip back to Jena and started searching for food. All I could find was some bread and sugar, a reminder to me of how bad things were in Jena and how necessary my food-gathering excursions were for our well-being.

When I reported to the Office of Work, the authorities received me with suspicion and addressed me with the unflattering familiar pronoun. After being in the West, where everyone used the polite second-person plural in addressing you, I found this Communist-introduced custom of equalizing everybody by reducing them all to the lowest denominator especially annoying. They told me that there were no high-paying jobs in Jena for people like me who had gone to the West; the only job for which I was qualified was digging ditches. Since I was already used to working hard, was well-fed, and loved the outdoors, I said, "Fine, I will dig ditches for a while."

I took out my wooden shoes and went to work as a ditch digger for an outfit that laid big pipes two meters under the ground. I enjoyed being outside, and I made a game of digging my trench in half the allotted time and then collecting apples and plums from nearby trees. I also enjoyed talking to my co-workers—several of whom were very intelligent and knowledgeable former Nazis and army officers who had been condemned to do manual labor for the rest of their lives. Some of them were very weak; I helped out if one of them had a bad day.

After two months, the people from the Office of Work ended my ditch-digging career and assigned me to Carl Zeiss again. I was still not allowed to work in my profession; I was assigned to do piecework on semi-automated lathe machines.

The department looked like a leftover from the early years of the industrial revolution. Each of the one hundred machines was driven by a very broad, oily-looking belt coming down from a moving labyrinth of transmission wheels suspended from the ceiling, which were themselves connected to the drive shaft of a huge motor. The whole room was oilier than a sardine box; even the air was suffocatingly saturated with lubricating oil and an oily cutting and cooling liquid. Today nobody would be willing to work in such an unhealthy environment, but at that time people would put up with far worse to earn money.

My job required great skill: I prepared cutting tools for each assignment and set them up correctly in a minimum of time. All set-up times had been fixed by stopwatch, so I had only so much time to complete each task. It was as if the human operators were merely gears meshing with other gears in a precisely timed, fast-moving mechanism. I was slow at first and could not earn the salt in the soup, but after six months I was as fast as the best. Then I set out to beat them. Everybody hated me: they were afraid I would cause a revision of the time standards. As I had never shown any enthusiasm for, or even polite acceptance of, the political forces in power, I could not even be held up as a model worker. (Almost everybody else despised the government, too.)

After nine months, the party bosses, urged by my worried colleagues, shipped me out of the department. I ended up working once more in my profession as a

Feinmechaniker, or technician, in a department where we built research microscopes and ophthalmological and measuring equipment. This department enjoyed great prestige. Some people had been working there for thirty-five or forty years. Having been part owners of the company before its expropriation and disassembly, they had always earned substantially more than workers outside the company. Some of the older mechanics still came to work in starched shirts and stiff collars. They told me that in the old days they made so much money they would throw it in the air and spike it with their walking sticks. I was happy to have the privilege of working with these skilled people and liked my work very much. But the thought of working in this department for the next forty years made me restless. My trip—all the freedom I had seen and felt—had fed both my bohemian nature and my ambition.

By this time I had already met Erna. I found her one day sitting in a meadow close to a lake where I had gone to swim. It was rather early in the summer, and the water still felt very cold. She was the only person I could see. She was sitting on a blanket eating cherries. I saw her from a distance and, as if magnetized, walked right up to her to ask her whether I could join her. I had never approached a person this way, but even from afar I was totally fascinated. If there were angels on earth, she was one. She was so beautiful, I thought, no Creator could have improved on this blonde creature. I was almost shaking and could hardly talk. I was glad when she invited me to sit down and offered me some of the cherries. I had seen many girls before, but her extraordinarily gentle nature and kind voice had such an effect on me that every goodness and charm I possessed was instantly unlocked.

This moment would remain engraved on my soul, to be reinforced every time I saw her thereafter. We dated for several months in what was to be the strangest love affair I have ever had—a relationship that was mostly platonic, yet very intense.

Erna worked as a housemaid for the wonderful Sandman family, whose house I still love today. Her employer was the marketing director of Carl Zeiss. They did not call her a maid, but *Haustochter,* or "daughter of the house." She was treated as a member of the family. She got all her clothes from the large collection Frau Sandman discarded. Erna always looked as if she had just stepped out of a movie scene. She was too lovely, too flawless in her appearance to be real: one could easily have believed she was made of the finest Meissen porcelain. Compared to her, I looked like Quasimodo. We spent wonderful Sundays together and always went for long walks. One time I was so foolish as to take her, when we were both in nice clothes, up to the forest I loved so much. She lost one of her high heels and had to walk back limping, looking not at all her usual graceful self. But she liked me.

We often went to the movies together. Those shown at this time were mostly Russian. Compared to the Hollywood productions we were used to, they were very crude and unskillful. Nobody was ever in black tie: everyone wore factory clothes or military uniforms. Characters addressed everyone else as "comrade," even if the other person was their lover or spouse. I remember one love story in which a man greets his girlfriend, "Comrade Natasha, I have brought you a present; it is a piece of steel from our new production."

At that time, I used to work with a man who also acted in the theater. He would often act for me love scenes from famous plays, and I would replay them for Erna. But I meant every word I borrowed that way, and made her believe me.

In a relationship between a man and a woman who

are in love, there always comes a time when their bodies melt together. But for us that took a very long time. It came one night when the Sandmans had gone on a trip to the West. Erna invited me into the house and gave me some chocolate. This was the first time I had ever been in such a grand house. The Sandmans had the most beautiful furniture, of a size I had never even imagined. I was so terribly impressed that I subsequently revised all my formerly limited objectives for my future. After we had made love together, which happened in a surprisingly natural way, I made the mistake of asking Erna how many men she had been with. Her shocking reply cracked my porcelain fantasy. She said, "You are the first man, my darling, but I was raped by five Russian soldiers when I was fourteen, when the Russians entered East Prussia." It was as if my heart had been ripped out of my chest, and we started crying uncontrollably.

A few weeks later, my brother was married in a village far from Jena. I attended the wedding ceremony and stayed a couple of weeks with the newlyweds. When I returned, I was stunned to find that Erna had a new boyfriend. He was a dental technician. I felt quite inferior to him, even though I did not know anything about that profession. I just imagined it to be much more prestigious than mine. When I encountered them in the Paradise Café, our favorite dancing place, I gathered all my courage, stepped up to the podium to sing *"Parle moi d'amour"* ("Speak to me of love") for her, in order to win her back. But just seconds into the song, the microphone broke down and neither she nor anybody else could hear me. I had lost her forever.

I could not get any food down for days. Whenever somebody spoke to me, I broke down in tears. I had learned both the elation and the pain love can bring and now understood how people could even commit suicide over it.

In different ways, losing Erna and having seen the Sandman's house and furniture contributed much to the way I lived and thought afterwards. I made up my mind to reach higher. At first, I had no idea what else I could be, but I began to go to evening school regularly. I also thought perhaps I might be able to capitalize on my good singing voice, and started to take voice lessons with the goal of becoming an opera singer. I also learned to play the piano so I could get all the notes right and also accompany my own singing. I even bought a piano.

Fortunately or unfortunately, the opera idea did not last long. My music teacher, who was gay, began to wear fewer and fewer articles of clothing when I went to him for my lessons. I was so terribly naive and innocent that I did not realize what was happening until he asked me one day before the lesson had started, whether I wanted to take a bath first. I remember running out of his house to the other end of town, in total shock. This was the end of my opera career, but I continued to sing and play in bands in my spare time.

During this period I encountered Blumenstein occasionally. He was very busy studying at the Workers and Farmers College of the University of Jena. He had become a socialist exemplar, a living testimonial to how a child from the working class could reach dizzying heights, even without a high school diploma in the new socialist paradise. He was promoted by all kinds of people as a *wunderkind*. To me he sounded as if he were reading from a book. Whether the topic was philosophy or electronics or some other scientific subject, he would leave us all scratching our heads after his second sentence. We who had known him earlier could not believe his transformation into a scientist. We started referring to him as "the professor." He kept trying to explain to me the most basic principles, but I failed miserably to make any sense of what he told me. He went on to become an assistant professor at the university. I felt that he had passed me by, and that was a terrible feeling indeed. I could not understand how somebody like me, with considerable technical prowess and a lot of common sense, could fail so badly in science; but I kept going to evening school anyway, taking courses in subjects that interested me or that I thought I would need to be better educated. I remember taking beginning algebra and philosophy there.

One day as I was seeing off a friend at the Westbahnhof railway station, I spied a person who caught my attention, a young lady wearing a hat and a half-veil over her face. At that time few women wore a hat and veil. Inspecting her face more closely, I was impressed by her great beauty and intrigued by her expression. I was so fascinated that I almost ran into her; I said hello and left my eyes fixed on hers until we started a conversation. She was from Bürgel, a small town close to Jena. I immediately recognized that she had a keen intellect and was quite taken by her. She told me she had come from Ilmenau after a visit to the School of Engineering there. I had heard that it was a great school, very hard to get into, with a very demanding curriculum.

I invited her to a little café nearby, and we got to know each other. I told her about my own situation and my ambitions, and how not knowing algebra meant that I could never get through the very tough entrance exams of any engineering school.

I liked this girl very much and saw that she liked me, too. There were still six weeks left before the entrance examinations at Ilmenau Engineering School, and she offered to give me intensive lessons in algebra that would enable me to pass the tests. I recognized that this was my chance, and I took her up on her offer. Although we liked each other very much, we spent many hours pushing formulas back and forth without either of us ever making any romantic advances. Later I always regretted not having slept with her. The only explanation I can give is that I was totally focused on taking advantage of this opportunity to advance my career.

The day came when we took a train the fifty kilometers to Ilmenau to take the entrance exam. I felt dwarfed and totally outclassed by the other applicants. Only one in four would make it into the school. I realized my chances were very slim. I did my best in the tests, but it was apparent to me I would not make it. And so it turned out: I failed to be accepted. I had lost my chance, not only to go to an excellent school, but also to live and study with this truly exceptional girl. If I had been able to perform better, the course of my life would have been determined for the next four years and maybe forever. My friend did get accepted and went on to Ilmenau, and I never saw her again. By that time of the year, schools had finished with their entrance examinations, so there was nowhere else I could apply. I ran around grinding my teeth for a while, trying to pull myself together and thinking about what to do next. Then I had the glorious idea of trying to get into another engineering school without an examination.

Jena had an engineering school, which was closely associated with Zeiss. I decided to make an eleventh-hour effort to get in. I had two advantages: first, I was an underprivileged "worker's child." Also, I had worked very hard at Zeiss, putting together instruments and testing them quicker and quicker, to the point where the old guys at the factory were very worried. I had applied to their environment Pepperl's work ethic and my own high-speed rhythm on the semi-automatic machines. The old-timers were prima donnas, used to working at a more comfortable pace. I particularly worried the party secretary, who was one of the slowest mechanics of them all. He had an interest in getting rid of me. This was significant, because without his strong recommendation and push I would go nowhere.

I explained to him that I had failed the test in Ilmenau, but that I would give anything to get an interview with the Dean of the School for Mechanics and Optics at Jena so I could leave the department and study engineering. He took the bait, talked to his party friends in the personnel department, and got me an interview. I talked to Herr Sieber, the dean of the school. I explained my situation very honestly to him: I had failed the test, but I was determined to become an engineer. Within the hour of the interview I talked myself up from no chance at all to a "maybe." This was one of the best sales jobs I ever did, and I could never have done it had I not had my experience selling to the farmers.

At Zeiss I kept working on the party secretary until he became my greatest promoter. He never mentioned my latent anti-government attitudes to anybody. Had he given even the slightest indication of these, my effort to obtain admission to the school would have been derailed immediately.

When I had left Herr Sieber, I knew that he liked me. One day before the semester started, he called me and said, "I will give you one semester to prove what you told me." I was in.

When school started in March, 1951, I met the forty-three other students who had successfully passed the exams. They were formidable, all better educated than I. But I got a great break. The school had decided to give an intensive refresher course covering the four years of mathematics everybody else had taken, just to make sure everybody was on board before even more difficult and advanced mathematics and physics were taught in the next semester. I was determined to take advantage of my luck. I studied as many hours as I could. I did not go to the movies, I did not participate in any sports, and I saw my beautiful and kind girlfriend, Gerda, only once a week for two or three hours. I burned the midnight oil every night, and I often doubted that I would make it. My mother asked me often, "Have you gone mad? You are killing yourself, why don't you quit?" But I had staying power; after six months I had reached the level of the other students. I was even elected head of the class and kept this position to the day when nineteen of us graduated with degrees in mechanical and optical engineering.

In spite of my successes as a student, I was still fragile and insecure. For my birthday Gerda and my mother gave me a little monkey puppet. They had even knitted clothes for it. Strangely, as soon as I saw it, I identified myself with that little monkey and hated it. For some reason, I thought I looked like him.

During summer break, my wanderlust hit again. I sold my piano to get some money and hitchhiked with my friend Günther to the Baltic. It was the first time I had seen the sea. After we arrived, I recall clearly the powerful impression it made on me. I loved the beautiful beaches with fine, white sand and the endless blue water reaching to the horizon. We found quarters in a barn where people were raising minks. We slept close to them every night. We were so exhausted from traveling that we slept even during the day in the big sand castles some girls had built. We would help eat their food, and when nothing was left, look for another castle to sleep in.

We walked the beaches for hours, right at the water's edge, splashing water with each step. Nature seemed untouched then. The water was crystal clear, and there were miles of lonely stretches along a protected area called Darth. On one walk we encountered a nude woman out there. She was very beautiful; Günther and I had to run into the frigid water to avoid embarrassment. We were young and easily aroused.

Whenever we encountered two girls, normally I would get the better-looking one. Eventually Günther complained, so I promised him the next time he could have the more beautiful one. When I was already sitting beside the more beautiful one in a beach basket, I remembered my promise. I traded him mine for the one he was with, and the girls cooperated with good humor.

Günther came home very late that night and told me he had a date with her the next day. When he went to pick her up at a street corner, he returned half an hour later very disappointed. He had not found her there.

I asked him, "Günther, did you wear your glasses?"

He said, "No, if I came wearing my fat glasses, she would run." He was indeed blinder than a bat.

I said, "Don't give up. Let's walk back, and I will see whether she is there." I spotted her, and told Günther to walk straight ahead thirty steps and he would run right into her. He did, and it was for good. They married two years later and are still together.

I, too, met someone there, a beautiful young girl named Christa. I knew immediately that this girl would play a major role in my life and that I would never forget her. But my usual approach of talking with a girl while we stared into each other's eyes would not work with her. Christa was wild and boyish, hiding her sublime femininity. We chased each other through the water and sand up into the dunes. One time she stumbled and fell behind one of the big beach grass-covered hills, and I fell on top of her. While I held her pinned down into the sand, I looked deeply into her mysterious gray-blue eyes. I almost kissed her, but did not because I thought it was unfair to do so while she was defenseless. When I returned to Jena, we wrote each other many letters, and later we spent a night together in

Leipzig. I remember her straw-blonde hair, her beautiful skin, and her perfect body. I needed and wanted her not only physically, but much more for who she seemed to be, a person radiating honesty. But we lived in different parts of East Germany, and it was too difficult to maintain a close relationship.

When I left for the Baltic Sea I did not want to look at books about engineering; instead I took along Goethe's *Faust*. As I read, I was so fascinated that I studied some sections over and over again. I was in awe of Goethe's work, as perhaps only a person without higher education can be. I don't know to what extent I understood everything I read without the benefit of a professor, but reading Goethe was a breakthrough I had never thought possible. For me it was the equivalent of opening a door to a famous person's private library and getting to the wisdom stored in the mountains of collected works.

Relatively untouched by life's great ups and downs, I still had much of a young person's natural sensitivity and felt deeply the tragedy that resulted for Dr. Faustus and Gretchen from his deal with the devil and his uncontrolled quest for happiness. Reading Goethe beside the beautiful sea lifted me into another sphere and made me forget the time in which we lived.

But when we returned to school, we found ourselves back in the other world, where little had changed since the war. The Communist government, remote-controlled from Moscow, offered only misery and the same type of propaganda as the Nazi government had. In some ways our lives had gotten worse because now everybody had to work harder to make the promised socialist economic miracle come about. Even we students had to go out into the fields to help the farmers on weekends, as did all the workers.

By now the Allies had implemented the Marshall Plan and were rebuilding West Germany's economy. Incredible amounts of money put everything in motion there. The Russians had no money and no desire to make such investments in Germany. They took everything we had for their own use while telling us that if we worked harder, things would get better. Taking foreign loans, they told us, would enslave us to foreign capital. We were supposed to pull ourselves out of the mud by our own hair. The Bolshevik propaganda machine came up with the most ridiculous statements about their war allies, the Americans, even claiming that one of the reasons we did not have enough to eat was that American airplanes had seeded billions of potato beetles into our fields. They also said that saboteurs had been sent over from the West to sabotage our plants and bridges—which justified the government in investigating anybody and everybody as a spy.

One day in school we learned another lesson: a person who did not read the newspapers every day could miss a change of policy from Moscow and end up in big trouble by spouting yesterday's dogma. The party secretary in our engineering class was astute, and 100 percent Communist. On the day when the government switched from insisting that no German should ever carry a weapon again to ordering that every German must learn to use a weapon to defend the country against the criminals of the West, our party secretary failed to read the paper. The wrong lines came pouring out of his mouth, and the incredible but inevitable result was that they fired their own man for expressing his pacifistic convictions. They didn't just fire him from his post as secretary of the party; he was also expelled from the school. The whole class rejoiced. The position of secretary was never filled because none of us qualified by being a member of the Communist party, or even a recognized believer; more than a few of us would have happily hanged all the Communists in East Germany had we had the opportunity to do so. (So would most of the East German population at this time.)

Everyone in our class simply wanted to become an engineer and hated the constant attempts to indoctrinate us. Studying Soviet ideological thought took up a lot of our valuable time. We had to read all the speeches made back in the Soviet Union and the repetitions of them made by our "leaders" in Berlin. All of them were miles long and would have taken up more space than all the books in our libraries. Another nuisance imposed upon us was the study of the Russian language. Most of us simply did not want to learn Russian, but we had to get good grades in it: bad grades in political science and Russian could get us expelled from school. So, while we were generally honest, cheating on Russian tests was considered okay. Three of our classmates came from German territories close to Russia and spoke and wrote Russian quite well. We made a deal with those three: they made several copies of their tests while they were writing them and distributed them secretly under the tables to the whole class. We then changed some of the endings on verbs and a few other things from right to wrong, to obtain test results that seemed to match our individual abilities. This cheating system worked for almost two years, until one day the student with the lowest grades happened by chance to correct the mistakes made in the original. He was the only one to get an "A." The teacher became very suspicious. When she called him to the blackboard and made him write, she quickly found out that after two years of Russian, he could not even write the Russian alphabet. She made the mistake of making a big fuss to the director. It backfired: he fired her for incompetence. We were free. We never got another Russian teacher, mainly because they were in short supply.

In 1952 the German Democratic Republic, as the East German state now called itself, organized a very big

"Youth Festival" in Berlin for the young people of all the socialist nations, with sports events and the usual parades demonstrating our solidarity with other nations, particularly the Soviet Union. Our entire student body had to go, and I, as head of our class, had responsibility for supervising the group. Shortly after we had been shown our quarters in Berlin—an empty factory building with straw on the ground to sleep on—I told my classmates that as far as I was concerned, they could do whatever they wanted; I would not see or report anything. Most of us went out to private houses, knocking on doors to see whether we could be taken in to sleep in real beds. The Berliners are very friendly and open people, and it was not long before all of us had become guests of families. We watched the sports events, especially the many fine soccer matches, and spent the rest of the time going across to West Berlin. This was easy to do: the trains of the subway system ran between the different sectors.

Thanks to the Youth Festival, which was intended to promote socialist pride, we had an opportunity to see with our own eyes what West Berlin was now like. With the help of American investment it had been made into a showcase. We experienced what we had only heard as rumors before—the free lifestyle and the total freedom of speech and the press the western nations had brought to West Germany. One could find virtually anything in the stores, so everybody had a great incentive to earn money with which to buy these goods. No loudspeaker systems screamed Soviet slogans and Russian music into everyone's ears. People went to listen to Harry James's trumpet solos, the big bands, and Elvis Presley, and if you did not want to listen, that was your privilege. We had a great time in Berlin, especially walking the *Kurfürstendamm* at night with all its lights on; if we had not had been focused on studying and becoming engineers, few of us would have returned to the East.

About one year later came the 17th of June, 1953. Every German should remember that day, but I don't know how many do, even though it is a German holiday. In the middle of a physics class somebody came in and said that the workers from the Northern Railway Station had marched into Jena's city center and were destroying the offices of the Communist Party with their coal shovels. Even though we had heard the day before something about demonstrations against the government going on in Berlin, we were stunned. This was happening in our city. Suddenly, everything we had been forced to swallow by the government came back up into our throats. It did not take long before we had all made the decision to participate at least passively by observing. We headed one by one to the center of town. A huge crowd was already on the move. Everything was totally spontaneous and unorganized. People were looking for persons they could accuse and punish for all the misery they had suffered over the past eight years. Nobody was to be found wearing a Party badge, though I am sure many party members and agents of the secret police were there in the crowd. Nobody seemed to care. Like the rest of the crowd, I felt my rage become more and more intense: I wanted revenge against the oppressors. Three men had hold of another by his clothes and were yelling: "This is the big party *schwein* that works with the Russians. He has put several of our colleagues into prison." Then they beat him badly. Nobody in the crowd objected. On the contrary, everybody started yelling, "Party swine, party swine!" And then, "Hang this party swine, hang him, hang him!" I don't know what got into me, but I was right there, next to the accused man, yelling the same words with the conviction of a judge. In this hysteria, regardless of whether this man was guilty or not, the crowds wanted to see blood. They wanted a release from the built-up tension of years of oppression and suffering. They wanted a visible, tangible demonstration of their new-found power.

No hanging occurred: no rope was to be had in the middle of the market square, and nobody could get through the pressing crowd to go fetch one. Then something happened that none of us had thought about during all these hours. We heard tanks, lots of tanks, and shooting. And then the Russian tanks drove right into the crowds and whole armies of soldiers started shooting above our heads. Still the crowd did not want to give up. It kept moving away from the tanks and soldiers, and then quickly, like water, flowed into the void left behind the tanks. People started building primitive barricades and pushing street cars into the path of the tanks. But what can an unarmed, unorganized crowd do against tanks and companies of soldiers? The only options we had were to disperse or be shot—and so, angrily, bitterly, we dispersed.. We then listened to the Western radio stations to hear what was happening elsewhere in East Germany. We learned that in the big cities, in Leipzig and Berlin, many people had actually been shot to death.

Within a few days all resistance against the authorities collapsed. It is still amazing to me that we actually thought the Western forces, the Americans, would come and help us. We had no idea how remote a possibility that was. The western powers did not even think of interfering; their intervention would have led to war. We wanted war, if it meant we would be liberated by the Americans.

Later I learned about the man we almost hanged. He was an agent of the secret police. Just before the three men had gotten hold of him, he had shot a fourteen-year-old boy through the head as people stormed the secret police headquarters to free the political prisoners jailed there.

10

When I finally graduated, having earned good grades, I remember that I walked down the street saying to myself repeatedly, "You are an engineer, man, you have made it to engineer, you are an engineer!" I had no idea of the difference between having an engineering education and actually being an engineer. But that is a common tailing among those who receive an education: very few people recognize the difference between having knowledge and being able to use it. Many educated people are unable to apply productively what they have learned; instead they use it to secure a position in which they stand in the way of the doers and achievers.

I immediately applied for, and obtained with no trouble, a job at Carl Zeiss. I was anxious to work hard, earn respect, and make some badly-needed money. I was looking forward as well to getting back out into nature on the weekends and holidays. Jena and its surroundings were ideally suited for that. I decided to take up photography seriously. It was such a joy to be free of all the pressure of the past three years, when I had had to study so hard all day and on through the middle of the night every day.

I had also broken up with my long-time girlfriend, Gerda—not because I did not love her anymore, but because I could not stand the pressure from her parents, who insisted that we get married immediately after I graduated. Though I had run out of ammunition to defend the status quo, I refused to surrender. The basis for my decision was very simple: I did not want to get married as young as my brother had.

Not too long after the breakup I went with a group of

young people on an overnight outing. It was early spring and nature was just awakening after the long, hard winter. We played games, hiked, cooked meals, made music, and sang songs. In this idyllic setting I fell in love with Erika. We had actually been studying together for two years, but she had been involved with another young man—and still was. She was so stunningly beautiful that at Zeiss, where she now worked as a designer, everybody referred to her as "Erika the beautiful." Erika played the harmonica, and I played the guitar. Singing the many love songs I knew, I had plenty of opportunity to win her over.

Our relationship became the most honest and idealistic, intense and powerful possible. Erika was the person who introduced me to classical music. She made me feel the beauty and majesty of Beethoven, Schubert, Brahms, and Mozart deep in my soul. She also took me to the theater; with her guidance I learned what opera can really give you. Even today, pieces like the overture to *Tannhäuser* can evoke exactly the same feelings and sensations I had when I heard them for the first time with her. We saw many plays I had previously just skimmed, and these now came alive for me. We lived for that period in another world, a sweet separate reality of heightened awareness.

All our acquaintances in Jena could easily see not only how much we loved each other, but also how we were deeply connected to the community. This is why, when we asked for permission to make a photographic expedition to the Alps, we were allowed to go. The authorities at Zeiss just knew we would come back. Our trip was the biggest undertaking of its kind anybody in Jena had seen. We had 350 pounds of luggage, most of it food. We had to pack heavily because we had no West German money and were not allowed to exchange any East German money to get it; we would have gotten very little in return for our money anyway, because the exchange rates then were close to ten East German Marks for one West German.

We went by train to Oberstdorf, my old stamping ground. Erika then bribed a bus driver with cigarettes to take us up to the very end of the Walserthal. There we loaded everything into a homemade two-wheeled cart and pulled that as far as we could. At that point we found a cable-winch system going up the mountain, which was used to transport milk and cheese. I climbed up to an *alm* (alpine pasture) and talked to the *Senn,* the herdsman who takes care of the cows for the summer and makes butter and cheese from their milk. He pulled our stuff up in exchange for a small can of meat. Then we carried the 350 pounds up above the tree line in three trips. There in a sunny meadow we found the most fantastic camping site, with panoramic views of the mountains, the meadows, and the forests. Everything seemed to be blooming and sprouting, making the meadows look like gigantic impressionistic flower carpets. On the sun sail above the tent we painted in large letters, "NIRVANA."

With the boxes we had brought I built a small but functional kitchen. I used tree bark to build an aqueduct which brought water from a nearby creek close to the tent. We would awake each morning to the most beautiful sights in nature, and start the day by bathing in the ice-cold waters that came tumbling down into the large pools in a nearby gully. We hiked and sunbathed, philosophized and made small talk. We got milk, butter, and cheese by visiting a nearby *Senn.* We walked barefoot whenever we could to prevent our shoes from getting soaked too much. Even though the ground was often freezing, we never had cold feet. We visited Pepperl down in the valley, and he and I renewed our deep friendship. Throughout the time Erika and I took a lot of photographs. No matter where we pointed the camera, we could not fail to get a worthwhile image.

It was the middle of August. The weather had been very good for the last two weeks. One evening it changed rapidly and produced a terrible thunderstorm. We had never seen such heavy rain. Thunder sounded almost continuously, deafeningly loud, and we saw lightning come down from the sky and continue like fireballs down the meadow next to our tent. It was very scary indeed, but finally, huddled together, we were able to fall asleep.

We must have slept a long time before I finally awoke. Since it was still dark outside, I continued lying, half dozing, half thinking, in my sleeping bag. Finally I could not wait anymore for the day to break, and got up. When I opened the tent, I could not believe my eyes. The sun, shining and bright, was directly above us. It was noontime, and our meadow had been transformed into a winter landscape. Snow had fallen knee-deep overnight, covering our tent—which had made it completely dark inside. The cows were standing in snow up to their bellies and could not find anything to eat. Walking around in the snow, we got soaking wet and cold. Two days after the storm, Pepperl hiked up to us and invited us to spend the last few days of our sojourn in his farmhouse. He told us that in our exposed campsite we could easily have been struck by the lightning and killed.

We wanted to use the snow to do some skiing. I had seen a big snow field high above on the mountain. Pepperl gave Erika and me skis but warned us to go carefully because there was the possibility of avalanches. We hiked up two hours to reach the snow. The steeply sloping snowfield was about 350 yards long. We climbed above it from the side and put on the skis. I thought it might be prudent if I went first by myself to test the snow, and left Erika watching me. I had just traversed about forty yards onto the slope when I heard a fine pop and felt a movement under me. In one quick motion, I let my ski poles fall, bent down and loosened my bindings, and laid down on my back, just as Pepperl had taught me in his mountaineering training years ago. Seconds later I was moving at great speed, as one with the entire snow mass. I went with it over a twenty-foot cliff, at which point everything broke up into big pieces, grinding and pressing against each other. I was still struggling to stay on top of this raging stream of ice when everything came to a stop in the trees, almost a quarter mile down the slope. I don't think Erika had any idea how close to death I had come. I had survived because I was on the upper part of the slope when the avalanche started, and because I came to a stop just before crashing into the trees. We searched for a while for the skis and poles, but we never found them.

Soon afterwards we said goodbye to Pepperl and his wife and children, and began hitch-hiking toward the East German border. We had no problem getting rides. Erika just had to stand by the roadside while I hid, and usually the first man to see her would stop to pick her up. These men were always very disappointed when I showed myself but rarely backed out of their offer to take her along.

When we got home, we had the pictures we had taken in the mountains developed; they turned out very well and were admired by everybody. The party bosses who had okayed the trip breathed sighs of relief that we had returned. And Erika and I gained more of their trust.

Becoming an engineer had changed my life. I was now a member of the so-called "technical intelligence" and had privileges not granted to the common worker. In a society in which everybody was supposed to be equal, we privileged employees received a much better lunch, served at tables with tablecloths and utensils, while the ordinary employees picked up their pots of soup themselves, paid the same price, and ate at benchlike tables. The supposedly classless Soviet society was actually intensely hierarchical, with visible class distinctions in every area of economic and social activity.

We heard from those of the deported specialists and scientists who had been able to return after six years in the Soviet Union that conditions there were even more unequal, and much worse than in East Germany. Nobody had any possessions, and even people of high rank lived with their families all crowded together in one room. In Jena, living space had also become a chronic problem. Most young people could not get a place to live; many newlyweds had to squeeze in with their parents for eight to ten years. Whether and when a couple decided to get married often depended upon whether they could obtain a space to live in.

With my new status, I felt that I had come a long way. I had even acquired a small motorcycle. With it, Erika and I were able to make many wonderful motorcycle tours, something all our acquaintances envied. But much of our free time was used up by the system. Often after work we had to participate in military training to prepare for a possible third world war. On many weekends we were trucked out into the countryside to do work the farmers could not or would not do. The pressure to join the party never ceased, and if you were not part of it, you could not advance to higher positions. The officials who did advance to positions of leadership regularly made decisions concerning things about which they knew absolutely nothing, going over the heads of people much better qualified to decide. The elections were all scams: you told the officials your name, and they handed you a ballot printed with the list of candidates chosen by the party. You were supposed to check "Yes" and drop the ballot in the box. True, there was one real election booth in a far corner of the polling place; but if you dared go into it to mark up your ballot, or even cross out a name, you could be sure that your name had been marked down and that "Big Brother" would be watching you more closely than ever. It seemed ridiculous that they required you to come and make your cross under *"die Einheitsliste"* (Unified List of Candidates). But if you did not show up to vote, they would come to your home and get you there. Maybe they felt that the chance to check everybody's temperature made their effort worthwhile.

I had long toyed with the idea of leaving, but I could not make up my mind. Too many things held me there: Jena was my home; anywhere else, I would be a stranger. All of us had an extremely strong sense of belonging to this little town in a way that I am sure is unimaginable for many Americans.

For some time I had been trying to get hold of some technical magazines from the West. When I finally did, I saw an ad for a job in Wintherthur, Switzerland. After thinking about it for a long time, I took a big chance by sending a letter. I never heard from that company. Not long after, I found an ad for a job in Malsch, near Karlsruhe, West Germany. I sent an application and received an answer in a plain envelope, saying they wanted to talk to me. By a strange coincidence, about the same time I also received a letter from an attorney in West Germany. It said that Frieder and I were the natural heirs to a piece of property and a house on it. This property was situated in Frankfurt am Main, and the lawyer said he would be happy to handle all the formalities to get it into our name; it would be necessary, though, for at least one of us to visit him and to sign the papers personally. This provided me with an excellent reason to obtain a travel permit to West Germany, and one was promptly granted to me. I was even allowed to make the trip with my motorcycle, as long as I left Erika behind. I didn't mind: I did not really know whether I would actually go through with abandoning my good job at Zeiss, my home town and its beautiful surroundings, my family, and all my friends just to be in West Germany. I felt I had not made many mistakes in my life thus far, and I wanted to explore the situation carefully.

It was very early in the morning and the fog was still lying thick over Jena and its little valleys when Erika and I walked down from my mother's apartment to the already packed motorcycle to say goodbye. I remember her saying, with tears pouring from her eyes, "Please come back, even if you decide it is good. Don't forget to pick me up." I replied simply, "Yes, I will."

Then I started the motor and took off toward an unknown future. I drove first to Goeschwitz, then on the autobahn up the Leutrathal. It was still early in the morning when, through the twilight and fog, I saw a large group of Soviet soldiers with trucks and equipment off to the side of the road. I passed them quickly, but I thought this scene made visible the sad situation East Germans were in. I promised myself that I would look at West Germany with wide-open eyes. Meanwhile, as I drove toward Frankfurt, I debated over and over whether I would be justified in leaving East Germany. After all, the government had, to a large extent, paid for my engineering education. But each time I came to the same conclusion: their investment in me did not entitle them to treat me as their slave.

Once in West Germany, I drove first to Wetzlar and talked to some people in the personnel department of Leitz, a big optical company like Zeiss. They weren't interested. They had absorbed too many people from Zeiss already—East German refugees were abundant in West Germany, and I must have made a very bad impression in my motorcycle outfit.

I was on my way to Frankfurt when a red light on the instrument panel of my motorcycle came on. I did not know exactly what it meant and decided to ignore it for awhile. The light kept flickering on and off. By now it was twilight. When I switched on the headlight, it was not as bright as usual, so I knew my battery was low and something was not working right. I switched the headlight off, driving with only the parking lights. I kept going, looking for a garage, and managed to reach the outskirts of Frankfurt before the engine quit.

Not knowing what else to do, I decided to push the bike to a gas station. I pushed for two or three hours. Nobody stopped for me. It was pitch dark, and I was totally exhausted when I finally halted on a tree-lined street with little villas on each side. As I stood there in the night, it started to rain. I had no West German money and could not pay for lodging anywhere. I began pushing the motorcycle again. I could not go far, and stopped to rest at a house with a separate garage on the street. The garage door stood open. I went in, sat down in a corner leaning against the wall, and fell asleep instantly.

Early in the morning I heard somebody come into the garage and leave again before I could open my eyes. About ten minutes later, the police, together with the owner of the house, appeared and began questioning me about what I was doing there. I explained my situation. The police asked the homeowner whether he wanted to press charges against me. Unreasonable though he was, the man finally had to say, "Let him go." But no one offered any assistance, and I knew then that in the West, you were really on your own. In a way this was good, but it was also scary.

My battery must have recovered overnight because my motorcycle started right away. I drove to Frankfurt and visited the people who were living in the house we were to inherit. It turned out not to be a real house, however, but an old railroad car that my deceased grandfather had converted into living quarters. I had never met any of my relatives on my father's side, nor even heard from them except when my father's sister in the United States sent a present one Christmas. The people living in my grandfather's railroad car appeared to be poorer than I. Although they were not my relatives, they had taken care of my grandfather until he died. I decided not to see the lawyer, and to leave matters as they were for the time being. I continued on to Karlsruhe and Malsch, and arrived later in the morning at the company where I was going to have my interview.

Not much later I shook hands with Herr Eckerle, the first capitalist I had ever met. His firm, which built special pumps, had about one hundred employees. We had a good and honest conversation; he liked me and offered me the job. He also offered to employ Erika, who had also studied engineering and worked as a designer at Zeiss. He even got my motorcycle repaired and gave me one hundred marks as reimbursement for my expenses, suggesting that I stay in West Germany now and let Erika come later.

But I thought Erika might not be able to leave if I did not return, and I had not quite made up my mind to leave Jena for good. Still, I saw here a great opportunity to get a foothold in the West, especially since I felt I could do a very good job for Herr Eckerle. I also liked the small village of Malsch very much. It was situated on the edge of the Black Forest with beautiful meadows, pine and fir forests, and deciduous trees all around. Like Jena, it had many fruit trees. In addition, Malsch was very close to the Rhine River. Karlsruhe, the nearby city, was also appealing.

Back on my motorcycle, I had driven through the

Black Forest and was close to Freudenstadt when it started to rain again. In Freudenstadt a car stopped so suddenly in front of me that I did not have enough time or distance to stop. I smashed against the car's rear end and flew all the way over its roof, rolling to a stop in front of the automobile. My suitcase and my belongings spilled all over the wet street, and my overturned motorcycle's engine raced and screamed. Four Americans got out of the car, helped me up, and picked up my things. They assured me it was not their fault, which I believed because the driver in back is always at fault in a rear-end collision. Besides, I had no intention of spending any time at the police station. My motorcycle was badly bent, and the headlight did not work, but it was still driveable. I continued, very carefully now, until I got to Pepperl in Oberstdorf. I wanted to see him and get his opinion about my decision. He listened to the details, said that it would do me good to experience a new environment, and wished me luck.

I drove back to the West German side of the border, where I stopped in a little grocery store. I wanted to change my Deutsche Marks West for East German money, but they said they had none. I knew I had to do something with the 100 Mark bill, so I hid it in a secret pocket of the little valise I carried. When I got to the East German border station, I was told to put my motorcycle in one of their garages for inspection and come into the office. An inspector asked point-blank to see my valise, went straight to the secret compartment as if he owned it, and found the money. I was then strip-searched, and left to wait several hours. Too late, it occurred to me that there might be a connection between the grocer in the West and the East German guards. How else could they have known to search only my valise and found the secret compartment so quickly?

Finally, two men in civilian clothes accompanied me

to an automobile and drove me to Gera, talking to me in a very friendly way. They had been ordered to bring me there, they said, because some higher official just wanted to talk with me. But I ended up at the Secret Police prison in Gera in a cell with four white walls, an iron door, one cold heating pipe going from the floor to the ceiling, and a small pot.

I spent three nights there, getting three small but tasty meals. I remained very hungry; I figured that cooking so well and giving you so little food must be part of their system to soften you up. I saw nobody. I heard nothing. After my ears had adjusted to the silence I could hear the guard walking by, even though he always wore big felt shoes. It was the first time I had seen a prison from the inside: it was shocking and calming at the same time. I felt insulated from all the responsibilities and pleasures of my outside life. My mind was blurred at first; then I started to assess my activities during the last few years, sorting them according to their degree of potential harmfulness. There was this job application, there were the many reactionary comments I had made against the state, and there were the foreign propaganda leaflets of the "Free Young World" I had found in the forest and never given to the authorities. So many things in my house could get me into trouble, and by now they would certainly have made a search of my room and workplace. I also worried about the money they had just found. Then I started thinking of things like my crossing to West Berlin during the Youth Festival activities and spending the night on *Kurfürstendamm.* When I thought back on my behavior during the uprising of June 17, 1953, I almost panicked. My trips to West Germany could be interpreted in some terrible way. Finally, though, I said to myself that what I had done did not matter very much, because in the eyes of this state *everyone* in the East had done enough to be found guilty of a political crime and put away.

On the fifth day, a guard opened the door and led me along a corridor and down some flights of stairs into a room with one desk and two chairs. An officer of the secret police who was sitting behind the desk invited me to sit on the chair across the room. He looked at me for a long time in silence. Then all of a sudden he said, "We know everything about you."

After another long pause, he asked, "What were you doing in the West?"

I explained to him about the inheritance.

"Tell me exactly what you did after you left Jena."

I had to tell him that story down to the minutest detail, and then again and again. Of course I left out the job interviews. Each time he said, "But we know there is more to it; tell me about that. Tell me about the one hundred marks."

I said they were from Pepperl.

"For what?"

"For the work I did a long time back."

"But there is more, we know it. What about the Americans? You can tell me, nothing is going to happen to you, just tell me. ... We are taking your motorcycle apart right now and will find everything. ... You have been in West Germany four times; we know you have been recruited by the CIA. ... You can improve your situation if you tell us everything." It went on like this all day. The next day he asked the same questions again and again so they could compare each small detail with what I'd already said.

I had really begun to worry I would never get out of there when on the seventh day I was brought into the same room. The same officer was there, but was now presenting a totally different personality. He invited me to sit down. He told me, "Listen, you have studied here, you have a great job, you really could help our state and repay what it has done for you. You are, by all measures, a very adventurous man, and I am sure you will find it exciting to work with us. We will even refund you your one hundred marks eventually. What we need from you is some information on what is going on at Zeiss. You are known to have a somewhat reactionary inclination, and this will be very helpful to gain everybody's trust and will make it possible for you to interview people we want to learn more about."

"Yes," I said, "This sounds adventurous and could indeed be very interesting."

I felt I could get out of there by being agreeable, and I was right. I got a code name, a password, and an order to meet my new liaison officer in the secret police in a week, at 3:00 p.m. close to the *Paradies Bahnhof* in the tunnel that led under the railroad track to the park. I was assured that it would be a mistake to run to the West or work for any of the foreign secret services because the East Germans knew all the Western agents and could punish anybody anywhere in the world. I was then given a pen and paper; they dictated to me a report containing precise troop movements of an Allied armored unit on maneuvers in Bavaria. I was forced to sign it with my full name and the code name that had just been assigned to me to make sure I was compromised if I later went to the West. Then I was driven back to the border, given my motorcycle, and told that my work with the secret police would open all kinds of doors for me, especially in my professional career. Then they wished me luck and shook my hand goodbye.

I knew intuitively that from here on, things would never be the same. Once a piece of porcelain is cracked, it no longer rings a true note when you tap it. I had seen and fallen into a netherworld that made the one in which I had lived seem obsolete and surreal, an unreachable fairy tale of youth. Even though I had not committed any criminal act—perhaps especially because I had not—I felt as if these prison walls had made me unclean, that from here on I would be different and never as virtuous and loving as in my past.

But I did not think even for one minute that I had a valid contract with the secret police, or that I had any obligation to work for a police state or spy on my colleagues, whether or not they wanted to leave for West Germany or had anything against this terroristic regime. I knew now that I had to leave immediately—there was no choice anymore. I had been caught in the gears of the machinery.

On my way back to Jena, I thought out a plan for extricating myself from East Germany before I had to meet the devil next week at three o'clock in the Paradise Park tunnel.

11

When I returned to Jena, I drove right to Erika's home. We went for a long walk, and I told her everything that had happened. Because communications were still slow in East Germany, we could assume that the Jena authorities had not yet been informed about my encounter with the secret police. We knew we had to leave before the information reached them. We also realized that we would have to leave behind, with keen regret, our few belongings: even though they might have looked pathetic to a West German, we valued them.

We agreed that I would try to get through in Berlin. We knew the subway trains were still running almost unhindered between East and West Berlin. But it would be too dangerous for the two of us to travel to Berlin together. There were eyes everywhere. When we bought tickets, we would have to show our identity cards at the ticket counters, and we had no good reason to go on such a trip together. So Erika and her stepfather would apply for a visa to attend the wedding of his daughter, Erika's half-sister, who lived in West Germany. Erika's stepfather was somewhat senile and would need someone to accompany him. His luggage would actually contain my clothes. As the wedding and Erika's West German half-sister had been made up, Erika could send her stepfather back home on the train as soon as they arrived in Frankfurt. He was left in the dark about all that.

The first steps of the plan worked very well. Erika got the visas; I had found a good reason to consult with a company in East Berlin. I confided in my boss, Herr Pichler, one of the people who had returned from the Soviet Union after six years of deportation. He managed to get me a travel order, the pass required to enter Berlin, which stated that I was on professional business visiting another factory for consultation.

Five days after I had been released at the border, I took Erika and her father to the railway station where I had once met the girl wearing the hat and veil. Erika looked incredibly beautiful. We said goodbye. I can never forget how her young, soft lips touched mine for an endless kiss, that tender touch incorporating all the feelings and sympathies we had for each other, and all our wishes and hopes for a safe journey and a future together. This kiss was one of those rare moments in life when everything comes together in one gentle expression of love.

Then she was gone, and I was free to think about what I had to do. I had already found two people who wanted to buy my motorcycle. After I sold it to one of them, the other got very upset, claiming I had promised it to him. He said he hoped fate would repay me for my bad faith.

I packed two bags with materials needed for the business trip, including quite a few of my technical books. Then I went through all my papers at home and burned everything that could be used against me should I get caught. I did the same thing at my desk at Zeiss. I felt a responsibility to leave all my affairs in order, as if I were going to commit suicide. I had loved working at Zeiss, and the company was like a second mother to me. It was extraordinarily hard to leave, and I don't think I could have uprooted myself from what I had in Jena if the encounter with the secret police had not taken place.

I said goodbye to a number of very close friends, and to my mother and grandmother, both of whom begged me to stay. But I thought, "Mothers always want you to stay: they want you to do nothing out of the ordinary, and worry if you just want to climb a mountain. Nature has designed them to be always dominated by their motherly instincts."

Now, though, I can better understand how difficult it must have been for my mother to see me leave. I was 26 years old and had lived with her for the last several years, after she had divorced her husband. I was her main support and source of entertainment. I felt very sad for her after I left her alone in the apartment.

At the last moment I decided to take one of my two radios with me, because I thought I could get a better start in the West if I could listen to their programs and get caught up on what was happening there. In the train I set it up in the luggage compartment, determined to deny that it was mine if we were subjected to inspection on the way. In fact, I expected the train to be stopped at any moment so that the police could come through to check identification and luggage. I was tired from the pressure of having had to make so many difficult decisions and the tremendous stress of winding up my life in Jena. I was worried about traveling into an uncertain future. Also, since Erika had decided to come with me, I would have no choice but to marry her soon and change my life forever. I was not so sure about that step, either.

When I arrived in Berlin, I decided I had better not carry all my luggage on the subway. Normally, commuters only travel with shopping bags and attaché cases. I checked half my belongings in the station's baggage check room. Then I took the subway to the West and checked the other half there. It all seemed very easy, a cinch. I had time to visit a doctor couple I had met long before on a climbing trip with Pepperl. They were very surprised to see me and invited me to stay, but I told them I still had to pick up the things I had stored. They warned me that it might not be such a good idea to go back to the eastern sector. But I needed the technical books with all their formulas and data to be able to work as an engineer and earn money. Replacing them all would have been very expensive and would have slowed my start in the West.

I took the subway back to East Berlin. It was very easy, and the life there looked very normal to me. I felt secure: I had done everything right. With my travel permit and my pass from Zeiss, my papers were in perfect order and I had a perfect reason to be here in Berlin. My appointment with the engineers of the company was scheduled for tomorrow morning. Of course, I never intended to show up there. Perhaps by that time I would be sitting in an airplane to Frankfurt, to the West.

When I arrived at the left-luggage counter, a crowd of people were checking things in and out. I thought that I should probably not wait around this station long enough to be noticed; I decided to pick up my bags later. I felt very hungry and had to find something to eat.

At that time, the biggest news in East Berlin was the construction of the *Stalin-Allee* (Stalin Avenue). Almost every day the East German press wrote about this gigantic building project that exemplified the glorious socialist future. It was East Germany's answer to the West's "show window West Berlin." Already half finished and occupied, *Stalin-Allee* indeed showed how the East German regime thought we were all going to live.

I had always wanted to see it and knew this would be my last chance. I decided to go there. When I arrived, I was very impressed. The *Stalin-Allee* itself was as wide as a soccer field. Along it were the biggest and brightest housing blocks I had ever seen. The stores on the street level had big show windows that were filled with merchandise we never saw in Jena. The whole scene was almost overwhelming.

I could not resist walking on that new street and staring at this brand-new socialist colossus. For a moment I thought, "Maybe there is a great future ahead for East Germany. Everything might get better—yet you are leaving." But I remembered the job I was supposed to do for the secret police, and that this entanglement left me no choice. Then my stomach reminded me again that I had not eaten anything since very early in the morning. I started looking for a restaurant or a bakery, but there was no place to buy anything to eat. None of those big display windows contained food.

I kept walking until I came to a store displaying cameras and binoculars. Curiously enough, the binoculars were the ones we made at Carl Zeiss. They contained the die castings that had given us problems in production and were, in fact, the pretext for my being in Berlin for a technical consultation.

I stopped in front of the window, and as I thought about how I had actually helped to fabricate these optical instruments, Pepperl Kessler's image came into my mind. He had always wanted a pair. Living high up on his mountain, he wanted to be able to see what was going on in the valley below, and also see who was on the way up to his farmhouse. I had just enough East German money left to purchase a pair of these binoculars. In the West, this money would be worth nothing anyway. I entered the store and bought the binoculars. To fill out the guarantee, the sales lady asked me for my identification. Seeing on my pass that I was from Jena, where they were made, she asked me why I was buying a pair here in Berlin. As she asked this reasonable question, I noticed her eyes shifting to a man right behind me, dressed in a whitish trench coat, who had been listening to our conversation.

In the next second, he presented a badge and I knew I had made a critical mistake. I was awake as never before; producing my documents calmly, I thought I did a great job of playing down the situation. But his suspicion had been aroused. There was no way of shaking this secret policeman, who kept his hand on the gun in the pocket of his trench coat.

Coming from the provinces and being an honest man, I was unaware of what had been going on in Berlin. There was a big black-market trade in certain East German products, which were smuggled to West Berlin for resale. The exchange rate of ten East German Marks for one West German Mark made it very profitable. I began to learn all this in the East Berlin office of the secret police, as I was being interrogated as a possible black marketeer. I learned more later that night when I talked with the people who were locked up in stand-up cells next to mine.

I remained in that first cell for two days. I got nothing to eat, little to drink, and no sleep. The police had arrested hundreds of people and were having a hard time interviewing all of them and sorting things out. I could sense that the officers were sympathetic towards me, because they really believed my story. They knew I was not one of the black marketeers they were after.

They had found my claim ticket from the luggage counter and picked up everything from the station. They could not figure out why, for a two-day trip, I was carrying enough technical books to design a whole factory, or why I had bought the binoculars. Even so, they almost let me go—which they later told me. There was only one more telephone call to make—to Zeiss, to see whether my travel document was authentic, or the forged kind used by the black marketeers when they came from other East German towns. That call sealed my fate. They did not call my boss or any other harmless civilian at Zeiss; instead, logically, they called their own secret police station at Zeiss, asking them to find out if my travel document was valid.

When the secret police in Jena learned I was in Berlin, they were stunned. They told Berlin not to release me under any circumstances; a car with three men was on its way. Eight hours later I was handcuffed and, with a secret agent on each side of me, driven back to where it had all started, the secret police prison in Gera.

I have never been able to talk about those three days. I did not even discuss them with Erika. I have always treated those days like a mirror I could never look into because it would be too painful to see myself appear so stupid. I had made an irreversible mistake that would bring me, Erika, and my family all the way to hell and back, a mistake that could never be erased or forgotten.

Back again in a cell in Gera, I felt as if a deep void had opened and swallowed up everything I had loved, everything I had learned and thought, and everything I had possessed. My lapse in judgment had rendered my entire previous world irrelevant. The secret police now had much more reason to believe I was a spy working for the West, and they had already concluded that I was about to betray them and the state by fleeing to the West. Only I knew I had never been a spy, for either side—I had never betrayed my friends and acquaintances, and I had never betrayed myself.

I realized this would become a time of trial, not so much from the outside as a constant trial and testing from within, a testing of my character, my perseverance, my resistance to severe pain, and my ability to deal with injustice and to make the best of each situation. I knew I was now as helpless as a grape thrown into a winepress. Nobody could help me, not even the interrogators, for they worked in this system, too. They were themselves locked inside the machine that had pulled me into its gears.

I paced back and forth in my cell trying to put things together, but I seemed to have lost my ability to concentrate. My mind was unable to focus on anything and kept imagining over and over a scene in which Erika was stopped at the border and interrogated by the police. For a long time I did not allow my imagination to go beyond that point. Visualizing how the Stasi might be treating her would have driven me insane.

Unfortunately, I had arrived early in the morning. After not having slept for three days, I was told by the guard I could not sleep until ten that evening. I knew then that these guys meant business. I barely made it through till the late evening, at which time the guard brought two blankets, one to put under me on the built-in wooden bench, and one to cover myself. There was no mattress or anything else in the room. When I awoke the next morning to this unreal reality, it was six o'clock. They had generously allowed me eight hours of sleep.

I spent the whole day imagining what might have happened to Erika. I felt great physical pain in the area of my heart and worried that my old heart problem was going to kill me. I wrote all kinds of scripts in my mind about how things had gone with her. Where is she now? She must be totally aloneout there in a foreign environment without her friends and family, and without me. She must have waited in desperation. Nobody could take such a shock without great pain. How had Eckerle received her? Would she be foolish enough to come back to East Germany and be imprisoned, too? The anguish of not knowing her fate and being separated from her was so intense and painful that it made me almost forget about my own situation.

Thinking continually of Erika and imagining the worst nearly drove me crazy. I could not eat a bite all day. Then at sleeping time, close to ten p.m., the guard came and brought me back to the interrogation room. After I sat down, the interrogator said in a very friendly voice, "Welcome back, prisoner," and then, without warning, threw an ashtray right into my face. I knew then what I would have to go through and how serious and brutal this was going to be. But I had no idea yet how low a person could be brought by isolation, mental torture, calculated starvation, and sleep deprivation. That very night, I got my first taste of the symptoms of sleep deprivation. After the interview ended at about 4 a.m., I was taken to my cell and fell asleep; at 6 a.m. I was awakened for another sixteen-hour day. When the guard caught me sitting on my bench dozing, I was punished by getting no food that day; and when the long, painful day had passed, I was again led to the interrogation cell and asked endlessly about the same story I had now told many, many times. Each time I showed the slightest sign of dozing, the interrogator would throw the ashtray, or kick me in the shins with his boots, or pour a carafe of cold water over me.

Later, in prison I heard about a former high-ranking officer of the Luftwaffe whom they had ordered to run through the room with his arms spread out, pretending he was flying an airplane, as kids do. He had refused at first, but after standing in a dark cell for twenty-four hours with water up to his chest, he did it; he said that at that point he would have done anything.

The night-interrogation routine went on for one week. In that short time I lost more than twenty pounds. I must have looked like the living dead; I certainly felt that way. I had been tortured with every imaginable threat and scenario, the most terrible one being that Erika had been captured at the border and been brought here to a cell just like mine. But she was having a much better time than I, making love to the interrogators. They kept returning to this theme because it hurt me the most. I did not believe she would behave that way, but the possibility of her having been captured was real, and having them talk in such a degrading way about a person I loved so deeply drove me nearly insane.

After that ordeal, I was left alone in my cell for three weeks. I saw and heard nothing except the guard opening the door when he brought my sparse rations and took out the pot. I was losing weight fast. To stretch the food, I would keep one slice of bread every day until it became very hard. Then I would eat it crumb by crumb, like candy. After I kept each crumb in my mouth for a long time, it tasted sugar-sweet and gave me a high.

Sometimes the sun projected a shadow of the window opening with its bars onto the floor or the wall. Each time I walked the three possible steps, it also cast my body's shadow into the window's image, making it look as if my body was pierced by the thick iron bars. This picture of myself nailed into the iron rods of my prison cell has never left my mind. Later I tried to paint pictures using this memory but was never able to fashion a painting that conveyed the feeling of despair I had as I walked my three steps.

I had a lot of time to think, but no great thoughts came to my mind. I compiled and went through lists of ways I might possibly get out and of people who might be able to help me. In the hell of those first weeks I tried to make contact with God but heard no answer. I was left to my own devices.

Routines keep most people from needing a psychiatrist. I realized I had to do something quickly to ward off insanity and so began packing my days full of activities. At first I tried to put my mental files in order. I started by making lists of all the things I had ever learned. Then I recited every poem I knew, every song I had ever sung, and so on. When I made the final account, however, I was somewhat disappointed that I had not learned more or retained more poems by heart. Even my knowledge of engineering seemed skimpy. I wondered why this was so. After reflecting on this a while, I decided that people process knowledge and learning as worms process the soil they eat: almost everything eaten goes straight through, and only a little gets absorbed. I knew thoughts like this were ridiculous, but I welcomed anything that would divert my mind from Erika.

I also had imaginary conversations with people like my friend Blumenstein. I tried to talk to him as he had talked to me and found it hilarious when I pretended he could not grasp my ideas, just as I had not understood his scientific deliberations. I also talked to Pepperl about great new mountains I wanted to climb with him. I even talked to Herman Buhl, whom I had met after he had conquered Nanga Parbat. I knew he had died some time after that, while climbing K-1 or K-2 in the Himalayas; but his being dead did not matter to me because at this point it was as if everybody on the outside had died.

I spent several weeks running the movie of my life in my head. Then I got the idea of trying to get back my job as a secret police informer, and escaping again. I bounced that idea around for days. At first it seemed an impossibility. Then I began to give it more credence: I reasoned that the secret police officials, like the priests of a religion, must be so convinced of the truth of their communist dogma that they would find it easy to believe that a convert was intelligent and sincere. When anyone converted, he could be accepted into the fold.

I decided to make the attempt and told the guard I needed to see the highest ranking officer in the prison. The guard brought me to him, and the officer asked what I wanted. I told him that I had had a lot of time to think and that all this was ridiculous. I was no criminal; I had simply panicked and tried to leave because they had scared me so much. He listened to me very carefully and then said, "Unfortunately you are already so deep in the state's mechanism that even I cannot do much for you now. Your value as an agent has also been compromised. Everybody in Jena knows that you have to be punished for having recruited your fiancée to go to the West. But after you have been sentenced and have gone to prison, you will have opportunities to work with us because our security service exists everywhere you will go, and you will be able to help us."

That was how I learned that I was to be put on trial. I had thought that I was experiencing the punishment and that I would be freed in a few weeks. Now I knew differently, though I had no idea what crime I could be charged with. In fact, from my perspective, the secret police itself had been the one committing crimes. They had taken me into custody at the border without evidence of wrong doing, put me into their prison, starved and threatened me, and forced me into their service. They had nothing on me, of that I was sure. I had kept to my story well, cleared my room and desk of all incriminating evidence, and avoided mentioning any actions or opinions that could be used against me. What legal reason could they have to put me on trial?

I also spent some time considering how I could possibly pacify the secret police or cooperate with them in order to spare myself the worst, but I could not imagine myself spying for them in prison. The same moral issues existed in prison as outside it. Besides, as a spy in prison I would either be punished for not supplying useful information, or valued for supplying it. In the latter instance, I would be set to further spying. I could conceivably be kept working forever as an informer in prison.

About this time another prisoner was put into my cell. Finally I had somebody to talk to. He had been on a visit from West Germany, he told me, and been arrested for voicing his opposition to the East German regime. He hoped to get out very quickly, and he offered to take any secret messages to the outside world and to Erika. He also taught me the special morse code system prisoners used to communicate with each other by knocking on pipes and walls. There were several longhand and shorthand variations, all used, even though everyone knew that communication of any kind was strictly forbidden, and that anyone caught engaging in it would be punished with less sleep and no food. Of course, you had no way of knowing who was exchanging knocks with you. The person who was asking you questions could be a Stasi employee.

The Christmas season came, and we shivered day and night in our very cold cell. I started to wonder how long I could survive like this. At first, I thought I would die in about one week, then two weeks, then four weeks. But I had already spent three months here, and I was still alive. They knew where the limits were. Nevertheless, I did not expect to make it much longer. The winter was getting severe, and I was down to skin and bones. My West German cellmate told me I did not look good and might not make it through the winter. The guards took him away and brought him back frequently; after weeks I concluded that he had only been brought in to spy on me. This turned out to be true, but I had woke up too late. I had already told him some harmful things, especially about my attitude toward the state, my hope that Erika had not returned, and my intention of going to the West if I ever got out of this place alive, because it had been too hard not to talk to somebody after so much isolation. Almost anybody could have become my friend at that point. Even the interrogators could have looked like friends to me if they had just called me out of my cell and talked to me. But after three more weeks, I found myself alone again. The West German was called out and never came back.

In February I was interrogated again. Everything was

written down and given to me to sign. The officer was nice and we got to be on friendly terms, almost joking with one another. I became so relaxed that I talked to him as I would have talked to my brother. I was, of course, hiding critical things from him, but I did not know what information the secret police were looking for, and I really needed somebody to talk to. I suspected they did not have any incriminating evidence against me. I also knew, though, that they did not really need any kind of evidence to convict a person, and that the presence or absence of evidence did not determine how long one would remain incarcerated. In fact, the officer told me, "You don't seem to realize that we, the secret police, are in power. We are even more powerful than the president of the German Democratic Republic. We don't report to him." This, I figured, could only mean that they reported to Moscow. It seemed logical. He also told me, "To be too honest is dumb." That was true; I had been quite naive. But at least I didn't have to worry about the other parts of my life in East Germany that I had kept secret.

Now it was deep winter and freezing cold. I was barely hanging on. I was just bones. I was waiting to get pneumonia and die. I often heard voices and sounds I knew did not exist. I saw flashes of things I knew were not there. I had studied the white walls thousands of times and knew every possible structure and apparent face their irregularities could suggest. I remembered the pictures I had seen of the people in concentration camps at the time they were liberated. I now looked like one of them. After five months, I was still wearing the same set of clothes I had worn to Berlin—the same socks, the same shoes, the same shirt. My shirt was now as shiny as leather. During the night, the sweat would evaporate from my socks, and in the morning they would be left so hard they would stand up like a pair of boots. Every morning I had to knead them with my hands to soften them. Otherwise, it was impossible to get them on my feet. In the five months I had been here, nothing had been washed, nor had I taken a shower. I had not read a single line during all this time, not even the things my interrogating officer had written. I had just signed everything he handed me, thinking, "Do what you want; I have no rights here." The picture of the living skeleton at the fence of the concentration camp kept entering my mind. Now I knew that even such an affecting photo could not record the length of a person's suffering. It could not make manifest the helpless deprivation and the fear of dying that had in part brought that person to look like this. Like a photo taken of a painting, it was only a resemblance, with the process remaining invisible.

I had some incredible dreams in this period. Once I dreamed I was having intercourse with a deer. Its large warm eyes looked into mine, while its hind legs were lying on my shoulders, holding my head. There was no violence, or resistance. I had as gentle and intimate a union as with a woman.

Another time I dreamed that I saw Adolf Hitler at a conference. I was surprised he was still alive. With him were many people I had only seen in newsreels. Officials from the East German government were among the crowd. The speaker's podium was right next to Pepperl's farm—close to the big cesspool into which I had dunked the crazy rooster. When the Führer spoke, he kept changing into Walter Ulbricht and back into himself over and over. This shifting speaker made the point that once one obtained prestige, one could never lose it again. Hitler kept referring to an earlier speech, at which he had fallen from the podium backwards into a cesspool. At that time, nobody had laughed; everybody had just expressed concern for the Führer's health. It appeared that great leaders could do no wrong.

After six months I was led out of my cell to see an attorney the state had named to defend me. I was surprised that my case required an attorney. I still thought I had done nothing. But the attorney told me, "Frankly, the state will not let you out of this, no matter what I say. You are probably looking at three years." I looked at him in disbelief. How could this possibly be? I told him my story, and he said, "You are being punished for having deserted the republic after you took advantage of the education it provided for you. You have talked another person of the technical intelligence into leaving the state. I will tell your story to the court, but if I open my mouth too much, I will be going to prison myself." With that he concluded our one-hour conversation; the next time I saw him was during the trial, just three days later.

I was very disturbed. The image of my father kept entering my mind. I was now just the age my father had been when he had shot all those people. I kept questioning myself: Was I actually going to trial for my father's deed? Was I suffering for what he had done so many years ago? I knew it was absurd to think this way. But the idea kept popping into my head again and again.

12

On the day of the trial I was handcuffed and brought to the courtroom—the only one I had ever seen, except in films. I had to sit in the box of the accused, with two policemen guarding me. My attorney was present but sat farther away. He did not talk to me. The judges I had seen in films had all worn black robes, so it surprised me when men wearing street clothes entered the room and walked behind the judges' bench. At that moment everybody rose. I remember only the face of the man who sat in the middle wearing a white ski sweater with a thick, high turtleneck. He looked rough, as if he had done a lot of hard labor. Only three people were in the audience section of the room. One was my mother. I could see she looked very worn, worried, and grieved. When I had entered the room, she had raised herself a bit and spoken my name. I felt great pity for her.

The man in the turtleneck began the proceedings. He stated that this trial involved a crime against the state, and that, for reasons of state security, the public would be excluded from the trial. I had always believed the state was part of the public. That was apparently not the case here. My mother and the two other people had to leave the courtroom.

There was no jury. A man spun out a story of my actions as an enemy of the state: my "boycott of the German Democratic Republic," a standard court charge against anyone who had stayed aloof from the system; my recruitment of a member of the technical intelligence to leave the state; and my attempt to export a valuable pair of binoculars to the West without a license—a violation of the law against "inner German trade." This was the first I had ever heard of such a law.

I must say my defense attorney surprised me. He tried to tell the court the truth. He began with the arrest on the border, my imprisonment and the pressure tactics used to scare me, and then began to describe my forced recruitment by the secret police. I had never heard my story from somebody else's lips before; I was so moved I broke out in tears. But before my attorney could finish, the man in the white pullover told him to sit down and shut his mouth. The trial went on, but I was not able to follow the last part. My mind had carried me far away, to the woman who was in the West waiting for me.

The trial lasted only twenty-five minutes. From the prosecuting attorney, I remember only, "Because of the severity of the crime, and taking into account that this man is very intelligent and has shown surprising criminal inclinations, I am asking for a term of eight years in a maximum security confinement." I first felt disbelief, then anger. I said again and again to myself, "You pigs, you pigs, you criminals, you criminals!"

I was taken out of the courtroom to wait until the judges decided. I asked the guard why the judges were not wearing robes. He said, "They have done away with that here. This is a Workers' and Farmers' Court. Judges doing this work have had only eight weeks of law training. All these men were miners before. They came from the radium mines." Later I found out that not only was this true, but these miners were conducting many trials a day, with all the sentences decided in advance of the trial by the Secret Police.

When I was called back to the courtroom, I learned I had been found guilty and sentenced to six years in a maximum security prison. It was as if I had been thrown off the edge of a world I belonged to and was falling into another world, gray and dark, one in which I could not exist, one from which I would never return. But after the judges had left, my attorney advised me that I should not take the length of the sentence too seriously. If I behaved well and did not cause any trouble, the chances were good that I would not have to serve the whole term. In a year or so I would be able to petition the President of the Republic for mercy. For now, he said, it was useless to do anything.

Despite his words of reassurance, the sentence left me totally devastated. I still could not accept as real the parody of justice that I had just witnessed. These men had so casually taken away such a long span of my life, and for nothing—years that belonged to me and would have been my best. I was only 26 years old, and could not imagine spending the next six years in prison. It was a time to consider suicide seriously, and I did, for many days. There was no way I could let Erika wait that long. I made up my mind to write her as soon as I was allowed. "Do not count on my ever getting out. Forget me."

Other than the total of six years, the most depressing and embarrassing aspect of this sentence was that after the common rubber stamp "Boycott of the East German government" they had given me five years for attempting to export a pair of binoculars to West Germany. Even though I knew this charge was absolutely ludicrous, it upset me deeply because it made it appear as if I had committed a crime against the State for money. I was afraid it would taint my image in Jena and in prison, where I was going to live for the next five years or maybe for the rest of my life.

I was taken to a large cell where all those who had been sentenced were being held. I was shocked to hear that all the others had sentences from ten years to life. I was the short-timer among them. But they told me it didn't matter how long or short one's sentence was, since the authorities could extend or shorten it at will.

Going to a real prison was a release of sorts, for life

inside was less severe than in the private facilities of the secret police. On the other hand, one carried the heavy burden of the certainty of spending a long time here. Still, everybody seemed as relieved as I was to be out of the hands of the Stasi. I was glad to be able to talk with other human beings after having spent six months in solitary confinement, though I had lost a lot of my ability to speak and could talk only in fractured sentences. It surprised me how much I enjoyed even just seeing the few tables and chairs in the cell. I liked their intricate forms. When somebody showed me a small piece of yellow paper with printed words on it, I was so happy to see the color and print I read it through several times.

We told each other our stories. After hearing them, I could not help thinking we were a rather pathetic lot. Nobody had attacked the system or shaken it with a heavy blow. All our offenses seemed ridiculous to me. Our convictions were for such things as writing down numbers of military vehicles and voicing opposition toward the state. One man, an engineer on a locomotive pulling trains to Moscow, had smuggled toothbrushes to the Russians. They gave him ten years for this. It was ludicrous.

A few days later a bunch of us were packed into a van and driven to Waldheim prison. There I had to exchange my civilian clothes for a dark blue prison uniform with yellow stripes down its side and across the middle of the hat. This day and the following one were the lowest point of my entire imprisonment. In that uniform I appeared no different than anybody else there, and felt worthless. It did not matter who we were or what we had done to earn our sentence—killed a wife, defended an innocent client, or done nothing at all. We were all the same, all criminals, all serving infinite time here.

That day the jailers led me into a cell in which nobody could have survived even two or three weeks. I thought this was the end for me. The building was not a regular prison with cells, but a 300-year-old structure originally built to store gunpowder. It was underground, with walls seven feet thick. Our cell had a light shaft extending eight or ten feet up to the surface, ending in a ten-by-ten-inch window with thick iron bars. The doors were made of heavy wooden beams held together with thick, hand-forged iron bands. Outside the door, immense iron bars secured with huge padlocks fit into recessed door frames to lock the doors. This small cell held four men and four masonry benches. We stared at one another's hunger-carved faces, and knew we could die here. Moisture dripped down from the ceiling and the walls, and it was very cold. There was no air to breathe. I did not even think of suicide any more because fate seemed to be taking care of everything. I remember very little about the three other people imprisoned with me. I was occupied with my own demise.

We remained there for ten days and, amazingly, caught neither pneumonia or tuberculosis. Perhaps small bacteria could not survive in such an environment. It was a miracle a human body could.

We did not realize that this was just a holding cell to condition us. Most prisoners received their initiation in this powder magazine. After ten days we were transferred to the maximum security cell block. This big building had 300 single cells, in each of which were kept seven prisoners. When we entered this facility, the most striking impression was the smell. It was a stink of the same sort as one encounters in the great ape cages of a zoo, but more penetrating. Each cell had six narrow shelves stacked side-by-side three high along one wall. An extra mattress was stored under one shelf; at night it was pulled out into the aisle for the seventh man. The shelves left a five-foot aisle, in which were one small table, two chairs, and one big clay pot with a lid.

With seven men in the cell, there was always one of

us using the pot. We got only one bucket of water a day. This was for all of us. We all had to brush our teeth and wash ourselves with this one bucket. To stretch the water a bit, we often used our morning coffee to brush our teeth. Once a week one of us had to scrub the cell, and eventually everyone got a turn to empty the pots of the entire ward. We were allowed to shower once a week, and had fifteen minutes every day to walk in the courtyard, weather permitting.

Two-thirds of the prisoners in this facility were enemies of the state, imprisoned for political crimes. The remainder were hard-core criminals. I soon found out I had one of the shortest terms of anyone in the prison. The others kept telling me they would be happy to trade places with me, even if they had to sit on razor blades the whole time of my short term.

I shared my first cell with two murderers, one robber, one safecracker, one counterfeiter, and a man who had raped and killed his 11- and 13-year-old daughters. In the beginning I was very scared at being locked up with them. Not knowing whether they would kill me or beat me up or attack me sexually during the night, I found it hard to sleep. But nothing of that sort happened. They turned out to be quite friendly. We got very skimpy food rations; seeing how thin and hungry I was, the safecracker pulled strings and got me two whole loaves of bread. I ate them to the last crumb the day I got them.

At some point I recalled looking during childhood at my mother's big medical book, which had fold-out pages showing all of a man's bones, muscles, blood vessels, and nerves. It also had a full page of human faces, classified by race and character. I remembered having wondered how the author had arrived at the face labeled as the criminal archetype. At that age I assumed anything contained in a book was absolutely true. It had occurred to me then that if one could thus identify criminals on sight, all criminals could be put in jail, and we would no longer need the police. Now, with a chance to test theory against reality, I looked closely at my cellmates and found that they did not resemble the degenerate shown in the book; moreover, they behaved about the same as "normal" people—the political prisoners, and even the people I had known in the world outside the walls.

I loved to listen to my cellmates' stories. It was like going to the movies, only the crimes they told about were real. The counterfeiter especially fascinated me, and I made him tell me his story again and again. He enjoyed doing it, acting out all the scenes precisely, moving about our small cell as if he were on stage.

He had been a robber since childhood, had been caught before, and had already spent nine years in prison. After his release, he had held up a credit union and used the money to finance a counterfeiting operation, working together with an engraver and a printer. What surprised me most was the incredible amount of work they had put into committing the holdup, buying the machinery, getting the right paper, executing the intricate work of copying all the details of the 100 Deutsche Mark bill onto the copper plates, and then printing and passing the counterfeit money. The police had caught his two partners after only three months, but he had managed to exchange one million Deutsche Marks over a span of three years. He bragged to us of having played cards with the police chief of a small village with his wanted picture posted on the wall above them. Unfortunately, he made the mistake of returning to his apartment to try to pick up some of his personal belongings. When he knocked, the door opened and he found himself staring down the barrel of a policeman's gun.

I really liked the safecracker, too. A very intelligent man, very reliable, and, in a way, very decent, he had learned everything about safes while working in a factory that made them. He considered what he was doing a profession requiring extraordinary skills, and looked down on the other prisoners, as a master craftsman on the outside would have looked down at a common laborer. When I was moved to another cell after four weeks, it was like saying goodbye to an old friend. Before I left, he asked me if I knew any place that had a big safe. I said I did not know of any and asked him, "Are you really going to do this again?"

He replied, "Certainly, this is my profession."

Every four or five weeks everyone would be moved to another cell and put with different inmates, for security reasons and also to allow the authorities to insert informers, to keep a watch on everybody. I came to welcome these changes. Once locked up with a new group, I would soon come to know each individual better than I had known anyone on the outside. There was no hiding in there. We were all stripped of the differentiations, the camouflage that the world permits in normal life. Here every movement, every word, every action, every mood could be read and assessed very easily. Over the next two years I learned more about people's behavior patterns and psychology than I could have learned in a lifetime of academic study.

It took a while before I recognized what incredible resources were here. Once I did, I made good use of them, and experiencing and learning from this universe of people helped me more than any other single thing to become successful later in my life. I met a greater variety of people than one ever meets in normal life. Some were a lot older than I, and a lot smarter, and we had time to examine a great many subjects philosophically. We tried to define the essence and the importance of love, friendship, existence, religion, death, and life itself. My cellmates taught me Latin and higher mathematics, as well as many things I had missed because I had grown up without a father. These people had been raised very differently than I. Most of the intellectuals had grown up in fine families and had attended the best schools. Quite a few had held high office or achieved distinguished careers. I enjoyed especially the time I spent with two lawyers from Berlin. Both of them had worked for the CIA and were in for life—as were the majority of the 2,000 here. We would philosophize endlessly, examining every subject to its finest branch and detail. These days of intense conversation opened new horizons for me and helped me to mature very quickly.

Reading and writing were usually forbidden, but there were some periods when we were allowed to read books, mainly classics and Russian literature. I read as many as I could and enjoyed every one. During one period, we even got newspapers. Most interesting articles, however, had been cut out, and the unprinted portions had been cut away: they did not want us to be informed or have anything to write on. We wrote anyway, secretly with toothpaste on slate.

Of course, there were also some very curious cases among us, but even these individuals broadened my views. Once I spent six weeks with a murderer who was a chemist and had dissolved his wife with acid in a bathtub. He was known to everybody as the walking encyclopedia because there was nothing this man did not know. He had a fiancée on the outside who wrote him the most beautiful love letters. He shared them with us; they were always a highlight.

Another murderer had preserved his wife in cans and had already eaten half of her before the authorities caught up with him. The rumor was that at his trial the judge asked him whether he had a last wish. He replied: "You would make me most happy if I could have another can of my wife."

There were people of nearly every profession—art

dealers, diamond merchants, jewelers, dispossessed factory owners, dispossessed real estate owners, dispossessed farmers who had been guilty of owning too much land, physicists, and others. There was even a fellow who had been the sales director of a large concern—a position comparable to that of Herr Sandman. When I was placed in a cell with him, he started pretending to be a waiter at the Hotel de Paris in Monte Carlo and recited for us endless appetizing menus he had memorized while staying there. Our mouths watered listening to him. He repeated this performance at every meal, and we ate our meals pretending we were eating all the fine things he was serving to us with his mind.

In reality, the food was awful. In the morning we got two small slices of bread with some margarine and jam. Usually at noon we got spinach or *graupen* (barley) soup. The spinach soup, which I had hated so much as a child, here was not really made of spinach but of the leaves of beets—normally fed to pigs. In the evening, again we received two slices of bread with little on them. We never got any fruits. To drink we received coffee brewed of toasted barley, tea, or water. We always said, "They know perfectly well on how little food a person can survive."

Prisoners who had been here a long time told us that in previous years it had been even worse, with many prisoners dying of starvation and disease. In those days, they said, when a cellmate died, they kept his body in the cell for up to a week in order to collect the extra ration. I did not know whether these stories were true, but we found them believable because for several years after the war the Russians had run the prisons, and we knew how many of the German soldiers captured by Soviet armies had died in their P.O.W. camps.

When our rations came, we took turns being first to choose. Otherwise people would fight to get the ones that looked larger. One man was especially ugly and ruthless.

We called him the *"Sauerschiss"* ("Sour Shit") because when he went to the pot, the stink was unbearable. Worse, he had to go to the pot each time the food came or while we were eating.

Occasionally a prisoner went from being perfectly normal one day to being totally crazy the next. I had never seen such transformations before. One man suddenly started to run into the wall, slamming it with his skull. At first we thought he just wanted to get attention; but he repeated it again and again, harder and harder, until he became a bloody mess and knocked himself unconscious. The guards took him to the insane asylum next door. Nobody ever returned from there. We suspected they were slowly killing the patients.

For some weeks I was in a different facility on the grounds, one in which about thirty men were housed in one large room. There I ran into Otto, the only inmate I had known on the outside. He was the son of my father's best friend, but he had lived in another part of Jena so I never had much contact with him. A superb mathematician, he had been given a term of ten years for having voiced his anti-government opinions. We spent a lot of time together playing chess and doing mathematical problems. He always appeared very collected and calm, but one morning we found him hanging from his bedpost. He had killed himself, using his underwear. Just the previous day he had seemed absolutely normal. His suicide shook me badly.

Almost everybody I met in prison expressed dissatisfaction with his previous life outside, and many wanted to change their professions. Everyone wanted to make more money, or at least make it more easily. Every second one was planning to open an ice cream parlor; that seemed to be the common prison dream for making the most out of nothing.

Every night of my entire stay, I dreamed of escape,

and I spent a substantial part of the daylight hours plotting and looking for opportunities. When I thought of escape, I often saw in my mind that long line of dirty shoes on the shelf back in the orphanage. Escaping from this prison was much harder. Many of the guards had spent years working in Nazi prisons. They knew every tactic in the book to prevent escapes. Nobody ever got away.

Once I developed a very good plan. I would use some wooden grills in the shower facility as a lever to bend the iron bars far enough apart to slip through. Then I would have to travel fifty yards along the top of a glass-spiked wall to reach the outside wall. I timed the guard changes and how often the light beam came over the top of the wall. I made special gloves and knee caps to protect my hands and knees from the glass. The escape could only be done with a partner. I looked around, but there were not many daredevils who were also trustworthy. The only one I could identify was one of the murderers. I talked to him and he said, "Yes, anytime. I have nothing to lose." I spent endless hours thinking whether it was justifiable to let a murderer escape just to free myself—and also about the trade-offs between an early escape and the risk of getting killed or crippled by a guard. The escape attempt never happened: a few days later, I was taken from my cell and moved to another prison named Torgau.

More modern than Waldheim, Torgau had served as a military prison during Hitler's time. I quickly found out why the Stasi had brought me there. They had a big engineering and design department right in the prison which, as fate would have it, did a lot of work for Carl Zeiss, Jena. I was employed right away and even earned a normal salary. Of course, I received only a minute part of it: the government kept the rest as reparations for the damage I had supposedly done to the state. The government also got most of what I kept, since I spent it on special food offered for sale to prisoners—milk, butter, marmalade, and even sausage. These foods were not considered luxuries. Everybody had a strong incentive to work: we needed the extra food to stay healthy.

Of course, even if I had been outside the walls, the government would still have taken most of my earnings, since it employed everyone, controlled all salaries, and set the prices (and quality) of any kind of merchandise one could buy. Money moved only from one pocket of the government to another.

In Waldheim prison I had been working as a watchmaker, but in Torgau I was able to apply myself to my profession as a designer, which did me considerable good psychologically. This technical work not only helped me in my later career but also led to my early release.

We prisoners lived in hope of something extraordinary happening. For example, we hypothesized that if the President of the German Democratic Republic, Walter Ulbricht, were to die, the event might trigger an amnesty. During the Hungarian Revolution in 1956, our hopes soared to incredible heights. We believed that American soldiers would soon look over the walls and all of us would be freed. But the West did nothing, and we soon collapsed back into our routine and our helplessness.

After the Hungarian Revolution, I dreamed of my release and developed plans for escaping to the West if the route through Berlin were to be sealed off completely. I thought of stealing a tank near the east-west border, and driving it across. Another of my plans involved crossing the Baltic Sea by kayak. I even designed a hot air balloon to use to float across the border. The major benefit of always having a plan to work on was that it generated hope. Curiously, other people had these same ideas, too; later one family actually did escape successfully in a hot air balloon.

Every day I exercised for half an hour in my cell. Sometimes I was even allowed to play volleyball, but this happened only on special occasions and depended on the mood of the director of the prison and on the political situation in general.

Several times my brother and my mother visited me at Waldheim. These visits were welcome, but each time I realized that we now lived in two worlds. Nobody on the outside could understand those of us in prison. In turn, the worries of the world out there had no relevance to our lives.

Each month a prisoner was allowed to write one letter of twenty lines, and to receive a like amount from relatives. We always tried to conceal double meanings in our sentences, but this was difficult, and only a few of our messages were understood on the outside. Erika eventually collected all her letters and mine and bound them into a nice book. I still have this book; I re-read all those letters recently and was very moved. They are a pen-and-ink monument to true love and suffering. I still feel very sorry for the young woman she was then.

My mother kept petitioning President Ulbricht for my release. Even after two years, her petitions were denied with the same form letter: "Because of the severity of the crime, a release after such a short time is out of the question."

The people from Zeiss visited the prison regularly to bring us design projects. Working gave me a way to keep my mind occupied. I used my expertise to advance myself into a position where it became apparent to Zeiss and the state that they would both benefit more if I were set free to work in Jena as before. I helped foster that belief in every way I could, though of course my plans were far different from theirs. Thanks to the inspiration and influence of my fellow prisoners, I was ready at a moment's notice to start my life over. I had clear ideas for my future: I was going to escape to West Germany soon after my release and rejoin Erika; I was going to earn back respect; I was going to have my trial and sentence declared unlawful and annulled; I was going to dress well after having had to wear this prison uniform; I was going to catch up and compensate for my time of deprivation and incarceration; I was going to learn at least one foreign language besides Russian; I was going to see and know the world better than most; and I was going to strive for financial independence.

For a man in prison—virtually pennyless and without significant connections—to set such lofty goals and then assure himself that he will meet all of them may seem the height of self-delusion. Yet in my later life I was able to accomplish all these objectives and more, beyond my wildest dreams. I think I realized even while I was in the East German prisons that my stretch of time there was a singularly valuable period of growth.

One day I was suddenly taken out of my cell and locked into a single cell. I thought it might mean that the secret police wanted to pump me for more information about myself or others. Or it might mean that I was about to be punished for having used paper from my job to write to the administration about inventions I had made for Zeiss, inventions that could be better exploited if I were working outside the prison. When I was informed that I was being released instead, I was totally astonished. Nobody in the prison, including myself, thought it could be true. We all thought it would turn out to be just another cruel joke: the day of my release was April 1, 1958, April Fool's Day.

13

When the prison doors opened and I stepped outside, it felt somewhat like leaving the theater after having seen a movie. After a deeply affecting movie one always needs at least a little time for introspection before one is ready to pick up other thoughts and return to one's own life. The experience I had just gone through, however, turned out to be powerful enough to haunt me for many years.

After I escaped to West Germany, and even later, when I went to live in the United States, I dreamed almost every night that I had made another unforgivable mistake and was back in a secret police cell. Night after night I relived the tortures I had experienced and the self-accusations following my arrest. Again and again I would ask myself the same question: "Why did you go back?" Also, seeing a person in uniform, no matter what organization it represented, could give me the shakes. Even a mailman or a conductor on a train or streetcar could affect me that way. But as I walked the three kilometers from the prison to the railroad station, I did not yet know of this impending after effect. I was just starting to wake up to a second life span, and being back on real earth was a great feeling. I had sorted out all the accounts from my previous life while I was in prison, and I had a clear picture of my life's balance sheet, with its balance of debits and credits, assets and liabilities.

As I walked, random thoughts entered and exited my mind. While I had worked out my long-term objectives, the suddenness of my release left me with no fixed ideas about what to do in the next few days or weeks. I did not know whether I should revisit the forests and meadows, or seek out my friends and my favorite places in Jena and reminisce, or find the people who had sentenced and tortured me and get revenge. I knew one thing for certain: I had done okay. I had not at any time given in to the secret police. I had not become an informer, as so many others had. I had behaved decently toward other prisoners. I was especially grateful to have come through this experience mentally and physically healthy. In return for a shortened sentence I could have volunteered to go to the uranium mines; thousands had not withstood that temptation and had lost their health or their lives there.

I knew also at this point that I could live my life virtually without a safety net. The four white walls of my cell had given me a new benchmark for how little one needed. To think and live one needs only minimal food and a layer or two of clothing. Not even shelter is an absolute requirement: if necessary, one can live as a vagabond, without walls, on the street. I had had the same secure feeling of self-sufficiency after I had finished my apprenticeship as a technician. I thought then that I would never again experience deprivation because I had a profession to fall back on. But now I no longer feared deprivation. I knew I could adjust my needs to a much lower level whenever circumstances might require it.

I had also lost the respect I had been conditioned to feel for and show to persons invested with authority. Before I went to prison, any person behind a desk could have intimidated me so much that I might have bent my head down like a Japanese or even clicked my heels. Now, a person had to earn my respect before I would trust, or defer in any way, to him or her. I had encountered in prison many once-mighty individuals in their pure human condition, stripped of their authority, without their desks and professional clothing, and I had witnessed how miserably most of them had behaved under duress.

I had entered prison a somewhat idealistic lover of forests, meadows, flowers, and butterflies. I emerged more experienced, tougher, more realistic, and full of ideas for the future. I felt more creative and more focused than ever before. And I was also a man fortunate enough to have no baggage: I had no wife, no kids, no home, and no possessions.

But I could not see how I would be able to take my place and resume my life among my friends in Jena. They had not lived under prison conditions, nor had they been exposed to all the kinds of people I had met and talked to. We now had very different experiences and premises on which to base our opinions. I could not even share with them the reasons I held the beliefs I did. The authorities had expressly forbidden me to discuss my experiences of the past three years. I feared I would have to sit silently and listen to naive and insulting statements, even when talking with good friends. Disobeying the order to be silent would be dangerous because I would not know who was spying for the secret police—taking the role that had been offered to me. For all these reasons I could not imagine ever living in Jena or working at Zeiss again.

When I reached the railway station at Torgau, I had my first beer. I then took a train to Leipzig, where I quickly realized that the situation in East Germany had gotten worse. An out-of-towner could not buy anything here: even food was hard to find. I had hoped to buy a new shirt, a jacket, and some pants so that I would not have to show up in Jena in the clothes I had been wearing when I was arrested. Somebody had attempted to clean them, but the effort had been futile: my cotton shirt still looked like leather, my socks were still stiff, and my jacket still had the same dirty, leathery collar.

Somehow I ended up in a place where people were dancing. The workers from a factory were celebrating, and they had beer. I just joined in, eventually finding myself seated beside a woman about my age. She became the very first person I talked to on the outside. After a while she asked me to dance, and I accepted. Because I felt somewhat clumsy, I said to her, "I have not danced in the last three years."

She looked at me and said, "I'll bet you just came out of prison." I was flabbergasted and wondered how she could have made such a good guess. The answer was simple: she, too, had just been released, after a five-year stay in Bautzen, a notorious maximum security prison in which many political prisoners had died. She told me she had been sentenced to ten years for her "boycott of the East German Republic." I knew that meant she had done nothing. Her opinions and attitudes had just been considered detrimental by the government.

It got late; there were no more trains leaving for Jena that night. We tried to find a hotel room for me, but there seemed to be none I could afford so I ended up sleeping on a couch in her friend's apartment. In the early morning I took the train to Jena. It felt strange to be back home. I seemed to have no connection with the town anymore. All the magic was gone.

I stayed only one day. I met with my half-sister Gudrun, who was only thirteen years old, but I could not talk to her very much. She was shy and not used to me, and probably in the dark about what had happened to me.

Then I visited my grandmother. She was quite reserved. I thought that might be because I had caused the family so much grief. Also, I was probably the only person she had ever known who had been in prison.

My mother was in the hospital, recovering from a cancer operation. They had taken a large growth out of her abdomen, and she was waiting for radiation treatment. She was happy to see me. I could see she had suffered almost more than I. I felt deeply sorry to have caused her so much grief. Quite early in our conversation I told her that I felt I had no choice but to leave immediately for the West. The Stasi had ordered me to report to them every three days. Since I was on probation, with three and a half years of my sentence remaining, they could use anything or even nothing to put me back in prison. I was scared. I did not think I could stand living in perpetual uncertainty, afraid of the consequences of any action.

She begged me to stay anyway; she needed me. She was also very afraid that I would get caught once more if I tried to escape. She did not want to lose me again. I told her I would go to Frieder and see what he thought, but that was only to comfort her. When I kissed her goodbye, I knew it might be forever.

I went back to the station and caught the next train to Oberheldrungen, the village where my brother lived. Frieder had visited me once in prison and without speaking a word had given me confidence that I could count on him, and encouragement to live through the imprisonment. As we had grown up, our relationship had changed. After I had begun living with my grandmother and had become independent of Frieder, he had become someone I wanted to emulate. Had I been healthier, I would have tried to be as good at sports as he was. Even though I myself had had no desire to become a *Jungvolk* leader, I had felt very proud that Frieder had advanced so far in the organization. As an apprentice at Zeiss, he had set very high performance standards, and I had tried to live up to them. I don't remember having had any arguments or differences of opinion with Frieder once we passed early childhood. He was always kind, always a good comrade to me, and I looked on him as the undisputed leader, showing the way.

In some ways I had not followed his lead, however. He had gotten married when he was twenty-one; his daughter was now nine. His wife had brought a little home and a nice garden into the marriage. Frieder had settled into the village. He was a teacher there, and one of the best players on a strong village soccer team.

When I arrived, my brother and my sister-in-law were very hospitable and supportive, but I could not calm down. I kept expecting that the police would pick me up at any moment. Maybe my release had been a mistake. Maybe I had said too much about what had happened behind the prison walls when I had run into several college friends in Jena. Also, the authorities had instructed me not to travel, and I had already violated my parole by being here. I knew I had to act quickly before they realized where I was and came after me.

After I went to bed that night, I became very restless. I was not able to fall asleep until early morning. Then I had a vivid and disturbing dream in which I was arrested for accidentally shooting a horse. In my dream I was taken at once to a court room. Everything there seemed to have been staged for my trial. I was made to stand facing the crowded visitors' area and a black cube, behind which sat the prosecutor.

The black-robed judges sat on a very high podium behind me. The head judge started to talk down to me. He asked me my name and date of birth, and then began reading various accusations against me. I thought this was odd because I knew that this was not the judge's job. I did not listen attentively because I already knew the story. I just kept staring into the stony face of the prosecutor.

I had been standing on the bank of a river. On the opposite side I had seen a big gray bull running by, a bull of the kind represented in a highly stylized terra cotta statue I had once seen. The bull disappeared quickly into the foggy distance. During that moment I had felt a strong desire to shoot and possess it. I was still wondering why when suddenly the bull came running back. I quickly fired a gun someone had handed me. But the moment the shot went off, I realized that it was not the bull but a beautiful Appaloosa horse instead. I knew I had hit it right under its head, but it was still standing unaffected when its owner came from out of nowhere and led it away.

The judge retold this story in many more words and concluded that the horse was innocent. When he had finished, the prosecutor stood up suddenly, at the same moment pulling a revolver and firing it at me. I did not feel any impact, but the loud noise made me sway as if I had been hit and was going to fall to the ground. The shot had been a blank.

Later they told me, "You have acted very well. We wish each of the accused could act like you. This revolver trick is a relatively new procedure to better demonstrate the guilt of the accused. Now the courtrooms are always crowded with visitors. The shot is fired at each trial because all the accused are guilty."

I could not sleep anymore after this dream. Though it was absurd, the dream was in some ways an accurate representation of justice in a Communist dictatorship.

I was now in such a state of panic that my brother offered to organize an escape from East Germany that very day. His plan turned out to be very simple. A friend of his would go to an out-of-town railroad station and purchase two tickets to Eberswalde, a place north of Berlin. His excuse for buying the tickets was that his parents were from Eberswalde. A reason was necessary because one was always asked at the ticket counter for identification and for one's reason for traveling. The ticket sellers were often directly connected to the secret police network.

Later that night Frieder's friend would drive us to another station closer to Berlin. As a truck driver he knew the roads well and the places where police inspections most frequently took place. We had selected

Eberswalde because to get there one had to go through Berlin, changing trains there. Once in East Berlin, we could catch the *S-Bahn* to West Berlin.

This improvised plan seemed very weak to me. I thought my brother was quite naive about its dangers and was taking a great risk. As for me, I knew that if I got caught, I would never see the light of day again. But my brother seemed quite confident, and I knew I could not spend another night in East Germany.

We dressed up like two auto mechanics, taking nothing at all with us. If we were asked why we were on the train, we would say that we were looking for work in Eberswalde, because we knew the place and it was beautiful. If we were caught, I would have to act quickly before I was handcuffed, while I still had a chance of escaping. I made up my mind that I would jump off the train if I had to, or even kill some policemen. If they shot me in the back, it would be better than committing suicide. I could not go back to prison: I had exhausted all my resources and could not suffer that way again.

Some hours later I was again on a train to Berlin. We had driven off after nightfall and everything had gone well so far. There were not many passengers on the train. We sat down with a group of soldiers. They were loud at first but eventually all fell asleep. My brother and I sat opposite each other. We remained silent. Each time the train came to a station, we watched the action on the platform carefully. We knew there were frequent patrols on trains to Berlin. Sometimes everybody had to leave the train and everybody and everything was inspected meticulously. Before we reached Berlin, the police came through the train once, asking for identification and tickets and checking all luggage. No one paid much attention to us. The train made many stops, and at each one Frieder and I sat on edge, expecting something to happen.

When we reached the outskirts of Berlin, another

patrol came through and checked everything more thoroughly. We pretended to be sleeping like the soldiers in our compartment; the guards woke us up and asked for our tickets. "Where are you traveling? Why are you traveling at night?"

We answered, "We are going to Eberswalde to look for work. We have no money to sleep in a hotel, and we have to be back the day after tomorrow."

It was about four o'clock in the morning, and the guards looked tired from checking the whole train at this early hour. We were relieved when they passed on to the next compartment; we fell back into our seats, looking at each other in silence, masking our emotions. I could now think of nothing but arriving in Berlin and getting through this ordeal. The time passed slowly, and the train stopped again and again. Closer to Berlin more people entered the train at each station until the train was packed, with some people even standing in our compartment and between our legs. I thought this was good; it would make it difficult for a patrol to work its way through to check anything. In our mechanics' clothes we now blended perfectly into the crowd of people who were obviously going to work.

It was early morning when we arrived at Berlin's *Ostbahnhof* (East Railway Station). The station was bustling, with people coming and going from the many trains. We walked as fast as everybody else, as if we knew where we were going. We knew there might be eyes searching so we did not stop to look for directions. Fortunately, out of the corners of our eyes we spotted signs pointing to the *S-Bahn* going to Bahnhof Zoo. I knew that station was in the West. I had taken the same train two-and-a-half years ago. Just five minutes later we were in the train heading in the direction of the zoo, holding our breaths, ready for anything. Minutes later we saw neon signs, more lights, and automobiles. We had entered West

Berlin. I prepared myself to jump out of the train should it for some reason pass through the Western Sector without stopping. But it stopped. We left the train and stood on western soil.

When Frieder and I walked out of that station, we knew we had entered a new world. We could not believe how many automobiles there were. My brother, especially, could not believe how different it was, for this was his first time in the West. In some respects it was frightening to see the bustling life here. We felt terribly poor and rather overwhelmed. Even in the two-and-a-half years since I had seen it, West Berlin had changed. People had become much more prosperous. I was now getting my first glimpse of my future life. I was being reborn here—emerging at the end of a metamorphosis that had begun when I had first said goodbye to my friends and family.

Now I had left my cocoon and was peering out into the unknown. The only thing I knew about this new life was that I would live it in the West. I did not know where or how, but I had a very good idea of who I was, what I had learned and absorbed, and what values I held. I did not know in what direction I should head, but I had the urge to unfold and dry my wings quickly and start flying.

I felt we had to celebrate this moment somehow. We had very little East German money with us, and it was almost worthless here. Alcohol was out. When I saw a concession stand with chocolates and a bunch of bananas hanging on its front, I bought two bananas to celebrate. I know this may sound silly, but for my brother and me it was not. No tropical fruits were available in the East, and I had not eaten a single piece of fruit during my entire time in prison. The bananas were therefore special and symbolic. I remember precisely the moment when we stood there with the banana peels hanging over our hands. I asked my brother then, "Why don't you go with me? We are both here; together we can go far. We can open a machine shop and a design bureau and make it."

But my brother, who had always been a leader, could not make the leap to the West and start from scratch. His wife, his daughter, his house, his garden, his students, his soccer team were all waiting for him.

When we had finished eating the bananas, it was time to say goodbye. We hugged for a very long time. Then he turned and climbed the stairs to take the train back home.

I knew it was goodbye forever. I would not see him again; the authorities would never let him out to visit me. The Russians would stay forever, and the East-West situation would never change.

14

Compared to the prisoners I had left behind, or even the people in the larger prison of East Germany, everybody in West Berlin looked vigorous and healthy. Their poise, their expressions, and their vitality showed that these people had something to live for. West Berlin was booming. The whole town was like a big display window. I felt this great freedom in my mind and soul, and I wanted to be a contributing part of it.

But not, I thought, in West Berlin. I wanted to get out of it as soon as possible. I could not feel secure here, in a city surrounded by East Germany: if the East Germans wanted you, they would find ways to bring you back.

I learned that escapees normally stayed in the huge refugee camps in Berlin for weeks or even months, during which time they were processed and interrogated by all the foreign secret services before being told where to settle. I did not want to waste my time this way. Besides, I could not stand the thought of living confined in a camp. I had just enough money for Pan Am's flight to Frankfurt, which had a generous twenty-dollar special fare for refugees. That same afternoon, April 3, 1958, I flew out of Berlin and across East Germany to Frankfurt.

I had already called Erika and told her I would arrive by train that night. She did not seem all that surprised to hear from me. Later I found out that her mother, whom I had visited two days earlier, had sent her a telegram: "Heinz released, Mother." She apparently had no idea that these three words could have prompted the secret police to put me under surveillance. I could have been arrested when I tried to leave Jena.

When the train arrived in Malsch, Erika was waiting. I was immediately struck by her great beauty and elegance. During the last two-and-a-half years I had become apprehensive about our relationship. In prison I had stared at her picture every day, studying her face over and over. In the process her image had become wooden and lifeless. I had told myself that this was only natural, that even if I had known the Mona Lisa but seen only her picture for two-and-a-half years, the same thing would have happened. A picture cannot maintain the power over you that a real person can, no matter how lovely an image it is. Over time, doubts had crept into my mind about what she was doing in the West without me, and whether she would be my woman forever, or whether she might react negatively to my intention to have an adventurous life.

When we first hugged and kissed, these doubts vanished. In that moment the magic reappeared; it was as if we had never been apart. Yet in reality, two different worlds were colliding. Erika, poised and impeccably dressed, had adapted completely to western life, while I was poorly dressed, without a dime in my pocket. My face showed the weight of East Germany's depressing atmosphere and the degrading years in prison; my skin, which had not seen the sun for years, looked gray. For Erika, seeing me like this must have been painful. I realized that I might be an unwelcome intruder into her new life in Malsch, and that she might well be wondering whether I could make it in the West. I wondered, too. I had always lived in situations in which I was taken care of. At home, my grandmother and mother fed me and took care of my clothes and my room, while the secure workplace at Zeiss was almost an extension of home. Even in prison, whatever little you had was given to you. You never had to make any decisions or choices.

Erika could not let me spend the night in the house

where she lived, so she put me up at the home of an architect friend who was on vacation. Waking up the next morning in my new surroundings worked a kind of magic. From the giant picture window I looked down over meadows and fruit yards, then far over the low plain of the Rhine River valley. I felt its beauty strongly. Alone in the house, I could stroll around and see and appreciate how wonderfully everything had been designed.

I went to the record player and put on a record. The very first sounds touched me more deeply than any music ever had before: it was Beethoven's *Egmont* Overture. At that moment I felt suddenly freed, ready and able to do great things. The future I had imagined in prison appeared attainable; it was almost as if I had already accomplished my objectives, because today I was living in the world I had imagined. Hearing the music, looking out at the beauty of the morning landscape from within this excellently designed house helped me clarify my objectives. I decided then that I would work as hard as necessary to enable myself to live in a house like this and perhaps feel the magic of this moment again.

In this mood I went for an interview with Herr Eckerle. He received me warmly and offered me the job in his quality control department that I had applied for two-and-a-half years ago. I told him my objectives had changed; now I would only consider a position as head of his design department. He was surprised and did his best to convince me it would be better if I started in a lesser position and worked my way up.

I later found out that he was more than just surprised: he confided to Erika that he thought I must have gone mad to demand such a position. Erika shared his opinion and told me so. But the design work I had done in prison had given me a chance to see what I could do, and to see how my skills measured up against those of others.

I was convinced I could do a good job as the head of his department and was not willing to settle for a lesser job. I felt a need to make up for lost time. I was not willing to spend time trapped in the small village of Malsch, waiting for Eckerle to promote me. I was willing to play hard ball. I felt that if I didn't get this job, I could always start a design firm in a city.

Having been confined for so long left me with a powerful compulsion to remove myself from any situation likely to restrict my freedom of action. This instinct to escape had made me feel that I had to leave East Germany immediately, and it had made it impossible for me to feel comfortable in West Berlin. This need for freedom extended beyond the physical and geographical into the social, economic, and emotional spheres of my life: anything likely to be confining triggered the same response—a tedious job, life in a small village, a marriage, any commitment that might block or limit possibilities for the future. This fear of confinement showed itself in many ways over the years: for example, I was never comfortable in a house with small rooms, or in rooms that had small windows or were windowless. I could not commit to any job in which climbing the ladder would mean I would have to give up freedom and flexibility. Similarly, I did not want to consider a woman as belonging to me, because I did not want to feel that I was caged in with her.

When Erika told me she had to go to Hannover for a few days to staff the Eckerle booth in an industrial show, I was stunned that she would separate herself from me so soon. After all, I had arrived only two days ago. Surely this gave her a reasonable excuse to cancel. I began to wonder.

Erika had moved me to the comfortable house of Frau Essig, the widow of a doctor. She was a very nice older lady with whom I got on well. I had telephoned Pepperl, who had told me I should come for a visit: he would pay for the trip. I decided to accept his invitation. It would be good to see and talk with him again. When I arrived, Pepperl and his family gave me a warm welcome. We went for long hikes into the mountains. In these beautiful surroundings I began finding myself and gaining my bearings again. At one point Pepperl said to me, "Listen, I should probably not tell you this, but when Erika was here last year, I came home late one night and found her sitting on the lap of one of my guests in the living room."

When Erika and I had both returned, we went for long walks through the lovely forests close to Malsch. We had time to talk, and found out slowly how different we had become. Erika was now a part of this community. She had become quite religious, and even asked me to talk to the priest in the village and to join in a religious procession he was organizing. I could not understand her feelings or thoughts. If I had not found God during the desperate times I had experienced, how would I find him now?

During one of these walks Erika said very suddenly, "I want to get married right away."

Feeling as if I still had one leg in prison, with no money, no job, nothing to wear, and no place to live, I told her that I needed more time to orient myself. "I am still in pieces," I said. "It would be reckless for us to take this step now. First we have to get used to each other, and I must find a job." I told her that I just needed a few weeks to overcome the huge changes I'd been through, and collect myself. I did not realize then how long it would take me to overcome the effects of my imprisonment and cease seeing the world and dealing with others in ways prison had taught me.

When I came home from that walk, I found a postcard Erika had sent to me from Hannover. On the surface, everything she had written sounded fine. But after reading it over several times, I felt that something was missing. I finally realized that the words had not come from her heart.

I now knew I did not want to work for the Eckerle Company. Because it was the only significant firm in Malsch and Erika already worked there, I was afraid Herr Ecklerle would have too much control over us. He knew our situation too well and could, if he wished, keep me down.

Erika and I decided that I should go to Karlsruhe, twenty kilometers from Malsch, buy some clothes, and look for work. When I had packed my few things, Erika accompanied me to the railway station. Everything seemed friendly and harmonious when we kissed goodbye, but I had the distinct feeling that she was counting on my staying longer than we had originally planned. I knew she was in somewhat of a quandary about what to do with me. We were now mismatched. I was poor, had strong opinions, was too adventurous, and was indifferent toward religion. When we said goodbye, I sensed that she was relieved.

In some ways I was relieved, too. Now I had only myself to worry about. I had to complete my metamorphosis, which required that I find a job and begin realizing the objectives I had established in prison. But the transition was not quite as I had imagined it would be, however. I was not really able to shake off the remnants of my pupa stage, dry my wings, and begin flying for almost ten years. During that time I had to find out who I could become, what I could do, what I really wanted to do, and how far I could go. Whenever I could see what came next on the track I was following, I kept asking myself whether that future was right for me, or whether I had to change my objectives to match new circumstances and new possibilities.

I began this period with 300 Deutsche Marks, which Erika had lent me. When I arrived in Karlsruhe, I went straight to a C+A department store. Within an hour I had bought a suit, a shirt, underwear, socks, a pair of shoes, and a tie. When I left the store, I looked and felt great.

I then walked to the Office of Work and luckily got a good lead. The Ritter AG, a dental equipment manufacturer with about 800 employees, needed a development engineer. I took the tram to Durlach and walked to the factory. It was 11 o'clock in the morning. By noon I had become an employee of the Ritter AG and had tucked a cash advance of 1,000 Deutsche Marks into my pocket.

Even in East Germany, 1,000 Deutsche Marks would have been a large sum for me; but this was the much more valuable *West Geld* (western currency). Near the railway station I had seen a "Rooms for Rent" office. I went there, and a nice girl helped me get a place very close to the Ritter AG, a comfortable room in the apartment of an older lady, for 50 Deutsche Marks a month.

I was settled. In the early evening I went to see an exciting cowboy movie, and I remember walking out feeling tall and strong. I was completely relaxed and energized. I slept soundly that night.

My luck continued the next day. When I reported for work, Director Schneider, the company's manager, called me into his office, and we had a long talk. He told me that he was very impressed with my background, that he needed a person like me to help achieve his own objectives. He explained that Ritter AG in Karlsruhe was a subsidiary of an American company in Rochester, New York; it produced top-of-the-line dental equipment designed in the United States. Director Schneider wanted to put together a stronger product line and had ideas for some improvements. He wanted me to be his assistant; I would be responsible for redesigning their existing products and inventing new accessories. I would have complete freedom to pursue new ideas and talk to any dentist in the country. My adrenaline was rushing. The whole situation seemed ideal for me.

The difficult moments in one's life, such as the points at which one drops an old life and starts a new one, often turn out to be the most interesting. I went to work with great elan. First, I studied the Ritter products thoroughly. I quickly saw that the American products were designed very differently from German ones, particularly the Zeiss scientific instruments with which I was familiar. I could see that the American products came from another world and were based on different concepts. By studying them I learned a lot about the mentality of Americans. Everything American looked somewhat improvised but was very functional. No time was spent on improving the appearance of any parts that were concealed by the housing; in German instruments, these parts would often be as finished as the visible ones. At Zeiss, perfection had been expected. For example, if a slip of your screwdriver damaged the screw slot, you were expected to replace the screw with a new one. You could be fired for failing to do so.

It did not take me long to come up with ideas for some critical improvements. We applied for three patents, and Director Schneider was so pleased that he doubled my rather low starting salary. I had no employee invention agreement, and the attorneys of the parent company in Rochester, New York, insisted on acquiring the patent rights for my designs. They offered me 4,000 Deutsche Marks. I accepted; at that time it was a large sum for me. I used the money to buy a new wardrobe, which I badly needed. I knew that in the West it was important to dress well and not appear poor. At that time I also felt a tremendous need to look good. The memory of grim prison uniforms remained strong.

With my new wardrobe I was able to assume the character of a prosperous, cultured young man who was at home in elevated social circles. I had no trouble keeping in character in public. My real self reappeared only when I returned to my tiny room, just as an actor who leaves the stage to return to his dressing room trades his stage character for his own less prepossessing self. Only many years later did I realize that most people in society behave in much the same way. When they are in public, they are on stage; at home, they live very differently.

I kept trying to see Erika. I took some trips out to Malsch, but she was either not home or had reasons why she could not see me. Later, Frau Essig, my former landlady there, let me know that Erika was seeing a young medical student. I then received a letter from Erika telling me she was convinced I could make it on my own and that it was better if each of us pursued our own opportunities.

Her letter surprised me. I thought that after all we had gone through, we could at least have remained friends. I received it in the hospital, where I was recuperating from a terrible bicycle accident. Her words sounded like ice: "Even though you have had this accident, I know you will be okay. I wish you good luck in your future life."

Fortunately, the medical staff took a liking to me because I was always joking. When the medics brought me to the hospital, I had made the doctor laugh. Instead of asking about the operation he was to perform, I had only wanted to know how good the food was. I was still a typical East German. They nicknamed me "Mr. Good Food."

A new friend, the girl from the "Room for Rent" office, visited me in the hospital. She was from Mittenwald and had eyes like twinkling little stars. She brought me a book, *Sinuhe the Egyptian* by Mika Waltari. I really enjoyed this historical novel, which diverted my thinking away from Erika into an ancient world.

After I left the hospital, I worked and lived even harder. I felt that I had to make up for lost time in both my professional and my personal life. Normal successes such as advancing slowly to better positions at work or marrying and having children seemed too small to compensate for the lost years. I was unwilling to settle in with one woman, or even to go on regular vacations. I searched constantly for ways to make money, to invent more things, to start a design office, or to start a business. In hindsight, I realize that I was still working with too little knowledge of how business functioned, and without any good leads.

I did come up with some small inventions, most of which were very simple, like a new nozzle for a shaving-cream dispenser. The one then on the market squeezed out a thick, sausage-like piece of cream that tended to fall off your face or your shaving brush. My little change distributed the shaving cream through many openings, so that the cream was foamier and did not fall off.

I offered my inventions to various companies, but the answers were always in the negative: "We've thought of this already," or, "We cannot pursue this now because we are busy with other things." I had not yet learned that one has to start a company to produce such devices, and then do the marketing and sales to achieve success. Now I realize that these little inventions might have brought in enough money for me to live on if I had had my act together. I did not.

I was also not ready to commit all I had to starting a company because I had an insatiable need for culture and a social life. I went out almost every night, usually to the theater, a concert, or the ballet. Almost anything musical would hold my attention at that time. On the nights I did not go out, I read late into the night, devouring books that had been forbidden or unavailable in East Germany.

I was also busy trying to find the ideal woman. After having been forced to live only with men for so long, male friendships did not much interest me. At first I was hindered by my desire to find a duplicate of Erika and replicate our love. My excess idealism and lack of understanding of how the world worked also hindered me as much in my efforts with women as in my efforts with inventions. On a bicycle tour through the Black Forest I met a group of French girls. They seemed to be just out of high school, but I immediately fell in love with one of them, an extremely pretty blonde. I was so much older and so shy that I didn't know how to approach her or single her out from the others, so I asked for all their names and gave them my address. The girl who spoke for the group wrote me several lovely letters. Unfortunately, she was not the one I desired. One day she surprised me by coming to Karlsruhe. She was very nice, and very intelligent but not that attractive to me. Looking back, I'm glad that at least I had enough tact not to insult her by asking for the address of her beautiful friend. I regretted very much, however, my inability to express the strong attraction I had felt for that girl. Still, this incident showed me that I was now freed from my relationship with Erika.

Meanwhile, at a tea dance I met a very elegant, smart-looking young lady from Zürich. I danced with her and asked for her address, but she did not want to give it to me. I wrote her anyway: I had obtained her name from her girlfriend, and I wrote her name on the envelope, adding everything I knew about her. The conscientious postal employees of Zürich delivered the letter to her. She wrote me back because she thought this was a miracle and God's will. She was very fine, very spoiled, very smart, and very attractive. We had many wonderful days together walking and talking. We admired each other, but I never even kissed her. When a friend told me, "If you marry this girl, you will work all your life just to pay for her clothes," I got scared and quit seeing her.

By then I had met Victoria at a swimming pool on the Rhine River. She, too, worked at the Ritter AG. She was a delightful blonde, and at twenty-one so beautiful that a photographer used her picture in his show window to advertise his work.

I bicycled home with her that day. Her father owned a factory which made all kinds of equipment for the building industry so I told Victoria about my inventions. She told her father about me, and so I was invited one Sunday for an automobile excursion. We ended up playing miniature golf. Her father wanted to hear about my inventions, but I had gotten carsick and felt so miserable I could hardly discuss anything with him.

Victoria kept pursuing me. I remember spending a memorable night dancing with her at a big event in the Black Forest Hall. Afterwards we sat together on a park bench with incredible fireworks illuminating the sky above us. We kissed and kissed, but when I asked her whether I could open her blouse, she said no. She started waiting for me at the gate every night; but when I asked her to go on a bicycle trip with me, she did not want to because she thought it more convenient to reach the same place by car. I knew this was true, but it was not the same. Besides, I had no car.

Then help came from the West German government. The courts had declared my trial in the East unlawful, had annulled my sentence, and had recognized me as a victim of Stalinism. The government sent me a check for more than 3,600 Deutsche Marks as compensation for my time in prison. Of course, no amount of money would have been enough to persuade me to endure what I had experienced. Nor did I think it was the responsibility of any Western government to compensate me. Nevertheless, I welcomed the money and used part of it to buy a brand-new Volkswagen. At this time there was a waiting period of three months to get a Volkswagen, but Director Schneider helped me get it right away. A Volkswagen was the best car I could safely drive to my job. The danger was not the traffic: Director Schneider drove an Opel, just one class higher, and driving a better car than my boss would have been unwise, just as it would have been a mistake to show up in the office with a more elegant tie than his. In Germany everyone had his place.

I had now been in the West six months and had come a long way indeed. Victoria loved me, more so now that I had a new Volkswagen, but I was starting to avoid her. I was still a little sour about her attitude toward the bicycle trip. I also worried that she might hinder my career at the company, figuring that if Director Schneider found out I was too involved with her, he might not advance my career. Only later did I realize how completely insane I was not to take the deal fate had handed me. Victoria would have rounded out my success very well. If I had married this beautiful, well-educated, well-mannered girl, I could have built a house with her father's help, connected myself to a family of entrepreneurs, formed my own company, raised children, and lived a respectable German life with her in security and comfort.

But I was just not ready for comfort and that kind of responsibility. I was still trying to do more, see more of the world, and taste more success before I settled down. I was in a very adventurous mood.

During the months that followed, I filled my life with women, but I was torn between my desire for love and my need to feel that I controlled each situation. My approach to women had changed—partly, I think, because I was still hurting from my break-up with Erika and did not want to get hurt again. I was still discriminating and loved each woman, but I slept with several just for enjoyment, avoiding serious involvement. I manipulated relationships to prevent women from becoming dependent or clinging to me. I thought later that by not getting more deeply involved, I missed out on something I valued: I knew these women only superficially, as if I were handling a nutshell without ever tasting the nut and getting its full flavor.

I don't think I was chauvinistic in my attitude toward women; I was only trying to catch up after my lost years and counting on my partners to make their own choices and decisions. It helped me to justify my actions and attitudes that women in the West were more materialistic than the girls I had known in the East and less motivated by their feelings.

By being close to so many, I learned a lot about women—what all of them had in common, and how their bodies and minds precipitated very different thoughts, feelings, and reactions than mine did. But in the end they remained mysterious and enigmatic to me, with a power to make me want them.

It was more than a year since I had last seen Erika when I received a letter from her telling me that she had quit her job in Malsch and intended to become a nun. I was floored. She said she was living in Karlsruhe, studying to be a nurse in preparation for her entry into the convent, and she asked whether we could meet.

To ease the pain of our break, I had forced myself to forget about her. It was hard for me to overcome my resentment and impossible to recover my old feelings for Erika, but I decided to visit her anyway. It turned out to be a very polite visit, with both of us carefully maintaining our distance.

I told her about the work I was doing, and that I might have to go to England for a while to learn English. Director Schneider had indicated to me that he had picked me to take over his position when he retired in three years. In the meantime, he wanted me to take over the management of production and services; English would be very helpful in that job, since I would be communicating with the Americans regularly.

Erika told me very little about herself. She simply said that her former job had not given her any chance to serve humanity, and she could do much more in that way in the convent, and also be closer to God. I could see she did not feel well. She looked pale, and I thought something major was bothering her. But she did not tell me. I offered her my help should she need it, but she was now in another world mentally, and was unreachable.

That night I drove to Malsch to speak with Frau Essig. She explained that Erika had been engaged to the medical student from Karlsruhe. Almost every night he had driven to Malsch on his motorcycle to see her. One night he was severely injured in an accident. If I remember correctly, his injuries had put him in a coma, he had lost a kidney, and he was likely to be crippled for life. After some time their engagement had ended. All this apparently had pushed Erika to retreat into her faith.

At the time I saw her, I did not have enough depth of experience to understand her. Her cool behavior toward me when I was hospitalized had hurt me so much that I could not sympathize much with her.

I had long conversations with my friend Uschi about Erika. Uschi was a chemist about my age, with a long-time boyfriend. We had met at the same swimming pool where I had met Victoria. I thought she was a true intellectual. She had an eclectically-furnished apartment that I loved. I wanted one like it; I tried to emulate other parts of her high-class lifestyle as well. We loved to philosophize together.

I could talk with Uschi about any subject, and she helped me enormously in learning to fit into Western society. We often talked about what I still needed from my past. Sorting through it reminded me of going through the scrap pile at Zeiss to find the valuable tools we had thrown there when the Russians made us dismantle the company.

Uschi even accompanied me on some shopping trips. I remember especially a wide-brimmed black hat, a Borsalino, she talked me into buying. Since it was the only such hat in town, it made me instantly recognizable from afar.

We took many walks together in the surrounding forests. Uschi often laughed about my prison psychosis and my paranoia about trusting others. I told her where I was going and what I thought I could accomplish. She did not really believe I would do all these things, but later in life she told me, "You are the only person I've ever met who is able to jump over his own shadow."

I had by that time, late in the summer of 1959, established myself fully in West Germany. I was not only a resident and a citizen, but I was also fully integrated in both appearance and manner of thinking. My provincial East German outlook had given way to a more nationalistic West German one. I had traveled through all of West Germany and had made long trips to Switzerland, Italy, and the Côte d'Azur. After prison all my senses were so heightened I could smell apples, oranges, and vegetables from far away, and could have bought them blindfolded. I could hear the slightest noise and the highest notes. Music could penetrate my depths and resonate powerfully in me, and paintings could provoke an excitement that lasted for days. When I traveled, I took in everything, hungrily and consciously. The world appeared beautiful to me.

Success in my profession provided great security and comfort. I had enough love in me to enjoy the world around me. I felt I had recovered from my imprisonment.

15

Director Schneider had challenged me to study English. I knew learning it in school would be a slow process, and I was afraid that classes would seriously cut into my pursuit of happiness in the evenings. Since I believed I also needed to see more of the world, I began to give thought to going to Rochester, New York, for six months or so, to work for the parent company there. That way I could simultaneously learn English and gain international experience.

When I suggested this to Director Schneider, he said, "No. If I send you there, you might not come back." His statement of concern caught my attention. When I inquired about salaries and conditions in the United States, I found that I would earn there twice what I was earning in Germany. Contemplating this, I decided I could have an ideal life there: if I lived frugally and worked for six months of the year, then I could travel for six months. I immediately wrote a letter to Ritter's main office in Rochester and applied for a job as a designer. They answered that they had no openings at this time.

I was now determined, however, to go to an English-speaking country to learn the language. Of course England was close by, but about this time neo-Nazis had begun vandalizing Jewish graveyards in England, and the newspapers reported that Germans had become unwelcome there. I decided that I would move instead to the United States. I did not realize what a big step I was about to take. It would have been simple just to get a visitor's visa, stay for a few months, and look at the country, but I could not afford to stay very long without earning money. This was the reason I decided to immigrate.

My father's sister Kati and her husband Reinhold lived in the United States. We had had very little contact with them, but once, years ago, they had sent us ten dollars for Christmas. At that time, ten dollars was a lot of money for us, so we had always assumed that she and her husband were well-off. I wrote to them, asking if they would vouch for me so I could get an immigration visa. They wrote that they were willing, and I went to work on the details. Soon I received a notice from the United States embassy in Frankfurt saying that I qualified for immigration and that I should fill out an application. At once I bought a small Langenscheidt English language course with records and a record player and began studying English day and night. Not long after, I drove to Frankfurt to pick up my visa.

While waiting in line I fell in love with the girl next to me. She seemed the kind of girl I had been trying to meet for years: poised, well-dressed, friendly, someone I trusted immediately, the perfect companion in America or the person who could make me stop the train, stay in Germany, and start a family. I think at this point I was desperately trying to hold on to someone representative of my homeland, a speaking piece of it.

Unfortunately, she was going to Los Angeles and I to New York; but we exchanged addresses and pledged to stay in contact and visit each other. Of course, I had no idea how big America was, and how hard it would be to reach Los Angeles from New York.

I also had no realization what kind of an adventure I was getting into, how hard I would have to fight for survival, and how much of myself I would have to give up—just about everything I had accomplished in my new life. Could learning a new language, seeing a new country, and embarking on a new adventure repay such costs?

Most of what I did know about the United States I

had absorbed from the movies. I had also had some brief contact with American soldiers in Jena, exposure to one product line designed in Rochester, New York, and one week living among Americans at CIA headquarters in Bad Homburg, being interrogated by German-speaking American officers. I had loved that week. I liked the American way of eating breakfast with bacon, eggs, and hashed brown potatoes, and I had enjoyed playing volleyball with American soldiers and Russian-speaking spies.

The CIA had wanted to send me back to East Germany as an American spy. "Would you go back to East Germany for us?" the American interrogator-recruiter had asked me.

"Yes, I would go," I said, "but only if you give me a tank to kill all those pigs."

I had added then that I might go to the United States someday. It was in the right direction—farther away from the Iron Curtain. That had happened only a year ago. I had never dreamed I would go to America so soon.

Once my visa arrived, I had to prepare to cast off—settle my personal affairs, quit my job, and buy a ticket to New York. I decided to fly from Frankfurt, even though most immigrants traveled by ship: flying was faster and cheaper.

And so I set my departure date: January 3, 1960. My resignation surprised Director Schneider. Though he was very disappointed, he remained good-humored about it, saying, "I do not worry one bit about you; you will make it. But if by chance you don't like it in America, you can come back."

For the remaining six weeks I concentrated on studying English. I had no time to waste. Since I was too poor to hire an English tutor, I called Uschi frequently to practice my pronunciation with her. I spoke the words as I thought I had heard them on the records. The results often sounded more like a nonsense language than English. It was fortunate that I had only a few weeks to study that way. I could have ruined my English totally.

Christmas came and went, and all my friends geared up to celebrate my departure on New Year's Eve. It turned out to be an unforgettable party. I did not bring a date, so I was paired with Ingrid, the sixteen-year-old daughter of the host. She had apparently admired me secretly, and she told me she wanted to go to America with me. We had lots of schnapps, beer, and wine. Somebody had received several large boxes of fireworks as a Christmas bonus from the fireworks factory where he worked. Toward midnight we went down to the street and lined up the empty wine and beer bottles on the sidewalk as sites from which to launch the heavy rockets into the sky. Ingrid had one of the big fire frogs, a bundle of firecrackers linked by fuses to make a chain of explosions. As she was lighting it, it started to go off, and she threw it away from her. As one explosion followed another, the fire frog jumped around wildly until it finally landed in a big box full of fireworks. Several of us ran to the box to save something, but it was too late. All the fireworks started to explode, and all hell broke loose. Everybody was in a state of panic. This incredible fire-noise-and-light show was more than adequate to celebrate the New Year and my leaving Germany as well, but many of us ended up with burns, holes burned in our clothes, or both.

This was not the last significant incident before I left Europe. At the party I had promised to take Ingrid for a drive to Baden-Baden and then up into the Black Forest on New Year's Day to pacify her for not taking her to America. We had a great time in Baden-Baden and in the Black Forest, though we never did see much of the forest because of the clouds and the intermittent snow. By the time we made our way back, it was late night and the road had gotten very icy. This winding section of the Black Forest *Hochstrasse* is cut into a steep slope, and I had to drive carefully. Fortunately, road crews had spread some sand. After a while I began to gain confidence, and accelerated out of each turn. Suddenly, we hit a curve that had not been sanded. The car spun out instantly, and I lost control. The car began turning in circles. Each time we neared the edge of the road, we expected to go down the cliff. Then we would hit one of the big side stones, bounce back toward the middle, and spin further down the road. Since I could do nothing, I just dived toward Ingrid and covered her with my body, waiting for the car to roll down the steep slope and smash into the trees. Instead, the turning and bouncing continued to be perfectly synchronized until, after eight collisions with the stone markers, the car came to a standstill. Miraculously, we were alive, undamaged, and still on the road. The car was in bad shape. All its fenders, bumpers, and doors were badly dented. After I bent some sheet metal away from the tires, the car turned out to be still drivable, and we made it home late that night. Ironically, I had already sold the car and was going to deliver it the next day. That was now out of the question. The car needed repairs, so I would get no money now, and far less than I had expected later. I would have to tighten my budget considerably for my first few weeks in America.

On the second of January I drove off from Karlsruhe to stay with a friend in Mainz. He kindly offered to have my car repaired and sell it for me. He and his wife also brought me to the airport the next day. All my belongings were packed in two suitcases. I checked in and said goodbye. My friend wished me luck, knowing better than I what I was getting myself into. Born in a village close to Jena, he had lived for three years in Canada. When he came back to pick up his bride-to-be, he had talked so much about Canada and the superiority of the Canadian ice hockey team that the government put him in prison for several years to keep him quiet. Eventually, he and his wife made it to West Germany, but she refused to leave Germany altogether, so they settled in Mainz.

I climbed the steps to the turboprop Super Constellation and thought myself lucky to have a window seat, even though it was right next to one of the big engines. But when they started up, the noise level was unbearable. Yet I was soon distracted by the German landscape moving by far below. I thought of Karl May's books and all the other adventure stories I had read. Flying to a new continent, where I would be able to see the places I had read about, was a big adventure, truly in the spirit of those books.

Thirty minutes into the flight, after I had settled comfortably into my seat, the plane started vibrating and shaking. Everybody around me looked startled and worried: this could not be normal. Minutes later the captain announced that we had engine problems and had started an emergency approach to Brussels. There was nothing to worry about, he said. We should remain calm and follow all emergency instructions carefully. When we landed in Brussels, the runway was lined with fire engines ready to help. Fortunately we didn't need them. We spent the night in a hotel paid for by the airline.

The next day we took off again. Over Ireland, the plane again began to shake and vibrate. We descended to a very low altitude and flew with reduced power over endless hedges and stone walls, passing farmhouses and sheep that, like us, were scared to death. This hair-raising, unexpected sightseeing trip was more of an adventure than I wanted. I was wishing strongly that I had taken a ship instead.

When we finally made it to Shannon Airport, fire engines were waiting for us once more. We spent another night in a hotel, but by now I was so scared about flying I could not sleep. How were we going to make it over the wide stretch of ocean to New York? Only the knowledge that the crew had to fly with us helped me keep my fears under control.

Of course I really had no choice in the matter: I had purchased a one-way ticket. I was relieved the next day when we boarded a new plane. We took off and everything went well. I was glad to sit by the window again where I could observe the engines closely. They seemed to be running perfectly, but the noise still bothered me. I diverted myself watching the clouds and the ocean.

The long flight gave me time to reflect. It had taken almost two years, but I felt I had fully recovered from being imprisoned. I think now that I was mistaken. My experiences in prison changed me forever. Certainly I have never been able to forget all that had happened. My porcelain was mended, but still broken. My feelings and sensitivity never again reached the warmth I had felt before being forced down into the dirt. They would always have to work through the protective layer of ice which I had developed to slow my emotions, and through which I now saw the world.

But as I reviewed my life since release, I felt I had met head-on all the challenges I had faced. I had adapted successfully to West German life. I had seen much of Germany and Europe and climbed all the mountains I wanted to. Pepperl and I had even climbed Mont Blanc, the highest mountain in Europe. I had seen enough theater and listened to enough classical music to make up for my lost years. I had walked the beaches of Italy and southern France, splashing my feet in the waters of the Mediterranean as I had so often imagined doing in prison. I had visited some of the great cities of antiquity and numerous museums. Now all these new images and experiences were firmly engraved in my memory, and I was carrying them to America.

Late in the evening, almost forty-eight hours after our takeoff from Frankfurt, I saw lights in the far distance. I was relieved. We were close to the North American continent. If the plane went down now I thought I might be able to swim there.

Not much later we began flying over land. We had made it! Soon we were flying over what I thought must be New York. The world below was cut into squares—streets dotted with millions of lights. I had never seen anything like this before. Each square held many single houses, and the whole seemed to go on endlessly. I was very impressed. My neighbor shouted to me over the engine noise, "Almost everybody in America has his own house."

After we landed, I realized immediately that I could not hear very well. The noise from the engines had affected my hearing. I could speak only a few words of English, and now I could not hear. This was going to be a big handicap. Nonetheless, we had to leave the plane, so I got up, straightened my tie and suit, slipped into my overcoat, and put on my Borsalino. Even though I had occupied one of the cheapest seats on the plane, I looked more like a first-class passenger. This was simply the German way of dressing for a big event. I wanted to make a good first impression and not disappoint my Uncle Reinhold and Aunt Kati. I had written them the time of my arrival, but the plane was almost two days late, so I doubted that they would be waiting for me at the airport.

When I left the immigration area, I searched the crowd but could not find anyone who looked like the photographs I had. Everyone seemed to be in a hurry.

I tried to ask directions, using the simple phrases I had learned from my record course: "Where is the luggage counter?"

"How far is it to Commack, Long Island?"

"Is there a bus to Commack?"

People seemed to understand me, but they spoke so fast that I could not understand their answers, even when I asked several different people the same question. Whatever they said went by in a blur, and sometimes I could only see their lips moving without hearing anything. After I managed to find my two suitcases, I stood there, holding all my belongings, not quite knowing what to do next, watching the crowd. They looked very different from the people I was used to seeing in Germany. There was an incredible mix of types; everyone seemed to be dressed very casually, or even poorly, for an important place like an airport. The policemen looked downright scary to me, as if they were used to handling rough crowds and criminals.

When I walked outside to the curb, I saw dozens of New York taxis and their drivers. I was shocked. Most of the taxis looked as if they had been borrowed from the junkyard, and their drivers looked so rough that I felt nervous about riding with them. I seemed to be the only dressed-up person here, and I realized that it had not been a good idea to travel like this. Just then a small man in a thick red hunting jacket, baggy pants, and a red lumberjack shirt with big checks approached me and asked, "Heinz?" It was my Uncle Reinhold. He had a rosy face. I could sense immediately that he was a kind, humble, and easy-going man. Compared to him, I was an arrogant, stiff-necked German driven to do things. Because I could not hear well, I could not understand most of what he was saying. We talked little as he led me to his jeep. It had a dilapidated canvas roof and plastic windows. "This is not a proper vehicle in which to drive in winter or pick up a foreign guest at the airport," I thought. Then I saw his tool box and power saw in the back, and his thermos and lunch box between our seats. My uncle was a carpenter and had come from his job.

We started to drive north on the Long Island Expressway. It was very cold; an icy wind kept whistling through the vehicle, and I had to hold my hat with one hand to prevent it from blowing off. My uncle started to tell me how he and Aunt Kati had immigrated to the United States in the late 1920s traveling by ship, how difficult it had been, how they had worked their way up slowly, and what this country was all about. I found it hard to follow what he was saying, and wished I were back in my warm, shiny Volkswagen on the way to a comfortable little café I used to frequent. This ride in the jeep of my American uncle was the oddest I had taken in my life. The contrasts between us could not have been more dramatic if we had been characters created by a scriptwriter.

We finally reached the little town of Commack. My uncle and aunt lived off a little country road in a small house he had built himself. My aunt received me with great friendliness, and we talked until early in the morning. I could not understand everything, so I had to guess a lot. They spoke rather broken German and often switched back and forth between that and English. They kept making German words into English ones and English words into German ones. My uncle and aunt were obviously not well educated, but there was nothing wrong with them—they were perfectly fine, normal and decent. I was the one who had to make big adjustments and cope with my own great handicaps.

When my uncle drove me around the next day, I realized how large the distances were in America. One could not walk to the shopping centers or supermarkets—one could go nowhere without a car. I also saw that my relatives lived in the boondocks.

I continued to be struck by how casual everything was. Many women went shopping with their hair messy or in curlers. I felt as if I had become stranded on a foreign planet. When I tried to watch television, I found I could not understand anything. My uncle saved me with his shortwave radio. I ended up spending my nights listening to German broadcasts in my tiny room beneath the roof, listening to every message as if each might be vital and relevant to my destiny in America. But nothing taking place in Germany was relevant to my life anymore. I had left the old world behind and had no money with which to go back. I had to face this completely new world starting from scratch.

16

I had thought that my hearing loss was a temporary effect of the plane flight, but it did not wear off. I had to see a doctor. My uncle warned me it would be expensive, and he was right. The doctor charged me ten dollars, a large amount for me at that time, just for removing ear wax that had accumulated over the years, which the noise and vibration of the airplane engines had packed onto my eardrums. I was embarrassed but tremendously relieved to be able to hear again.

Once I could hear, I studied English as intensively as I could. It discouraged me that I was communicating at about the level of a five-year-old. I realized it would take me years before I could speak and think in English, as the Americans did.

My uncle and aunt helped me learn by sharing their insights about how people lived in the United States. My head became saturated with new English words, information, and impressions. Sometimes I got headaches trying to read every road sign and advertisement I saw.

I kept noticing ways in which America was different from Europe. The most visible differences concerned how people lived, dressed, and ate. In America people shopped cost-consciously for the least expensive goods. They wanted several different articles of clothing of each kind, and they changed clothes every day. In Germany people were more quality-oriented. A person would buy one high-quality item, like my Borsalino hat, and wear it with great pride forever.

The first time I went to the beach I discovered another difference in American clothing. My skimpy German swimming trunks attracted a lot of attention. I was almost attacked, and people yelled as I passed, "Thanks for showing." I was definitely in a more puritanical society. (Even now, this difference in standards is noticeable. When I am at a beach in Europe, perhaps three-quarters of the people there are nude, and no one makes any fuss about it. In much of the United States, people still act scandalized when a woman feeds her hungry baby.)

At that time, American cars were enormous, with huge finned fenders. European cars were smaller, more conservative, more streamlined. What really amazed me, however, was that one could buy a new house in America for about $12,000. In Germany such a house would have cost five times as much. Nor were savings needed. Most Americans bought everything on credit—even homes were bought without a down payment. Germans saved a lifetime to realize their dreams; Americans fulfilled them instantly.

American food, too, differed greatly from European. I found its high fat content hard on my stomach. Eating, like almost everything in America, seemed to be designed for speed, with instant coffee, instant foods from the refrigerator, and instant meals ordered and eaten in fast-food restaurants, or standing in line, or even in the car. Services were organized for speed, too. Shops returned laundry and dry cleaning in one day, or even just hours.

I had lived under planned economies, in which everything was regulated. Here, five-year plans did not exist, and the economy ran itself according to supply and demand. Everything seemed unorganized, decentralized, and improvised. Yet goods and services somehow found their way to whoever needed them, or at least whoever had been persuaded that he did. Nobody in authority asked how much profit a vendor made on something. The customers' questions were, "What can I get it for? And is this the lowest price in the marketplace?"

People who went to the bank to apply for a loan wore

the clothes they would wear to the supermarket. There was no need to dress up, as one did in Germany, because bank loans were based on one's earning power and assets, not on one's appearance. There was pressure to acquire as much as one's neighbor had, but nobody felt the need to signal his status by wearing formal clothes and an aloof demeanor.

Americans were incredibly polite and friendly. I was surprised how easily one could make contact with them and how they overlooked my poor English and treated me right away like a countryman. Their friendly ways often confused me, though. Many would tell me to come to their homes, but few invitations actually materialized. Eventually I learned that invitations of this kind indicated that people liked me and wanted to be friendly, but that because the offer was made without time and date, it meant that they were not actually extending an invitation for me to visit them in their home. Later, even good friends would suggest that we meet in places many thousands of miles away in foreign countries rather than just get together in the town where we both lived.

Americans also tended to compliment you on everything—a nice car, a nice house, a nice shirt. Then one had to say, as they did, "Thank you," for each compliment, and receive each time the response, "You're welcome." I thought about this custom and concluded that it was an easy and harmless way to make somebody feel good. I had a dream once about this politeness, in which I was shown a 150-year-old house by its owner, who hardly could have been its architect. I told him, "I love the architecture. You have done a very nice job."

I spoke English with a heavy German accent. Americans who had any German ancestry would immediately say, "I am German, too," even if only one grandfather had been German and they themselves knew not even one German word. I was surprised at how frequently

Americans mentioned how much blood of each nationality they had, emphasizing one-eighth, one-quarter, or one-half German, French, Irish, American Indian, or whatever. I was not used to such information from Germans, but I remembered how Hitler had drawn legal lines condemning any person who was less than half German, particularly those with any fraction of Jewish blood.

Americans seemed to me far less conservative than Europeans. I thought one reason might be that every American seemed to carry a bit of the entrepreneurial, adventurous spirit of the country's settlers, who had rejected an orderly hierarchical system by leaving Europe. Many of their descendants still exhibited the frontier dweller's attitude when talking to a new acquaintance: "I am starting out new here. What do you do? What is your name? Where are you from? How was it there? How did you make it? Maybe I can do that, too." People seemed constantly to be searching for a spot in the system where they could identify an opportunity and supply a need. Yet there was a friendly welcome for everybody, as if to say, "This is a great country, this is a big country with enough room and work for everybody. I wish you good luck, nice meeting you. I'm sorry, but I have to hurry off to find my own luck."

The attitude toward culture was very different from that in Europe. An American might have explained it by saying, "We have everything Europe has: opera, symphony, ballet, theater, Broadway plays, and a repertoire of so many new varieties of music they could keep you excited forever. It is all up to you, though, to enjoy or ignore. There is no obligation to listen or participate. There will be no embarrassment or stigma if you didn't attend a performance, or didn't like what you saw or heard."

The price for the freedom to follow one's own preferences was that one had to make room for whatever satisfied everybody else. Preferences emerged from the backgrounds of the many races and nationalities brought together, the many religions, the many customs and tastes, and the never-ending variety of opinions, expectations, and expressions. One had to make room in oneself, too. Whoever you were and wherever you came from, whether you wanted it to happen or not, you would be partially homogenized, losing some of your cultural identity in America's big melting pot. It was left to you which ingredients and how much of each you would in turn take back from the pot to incorporate and use, but some would attach themselves to you whether you liked it or not. Did I like it? I was not so sure. I did not want to give up part of myself, or change: I just wanted to learn and see. And somehow I had to earn money, even if it was just enough for a ticket back to Germany.

My uncle and I went on the road to find a job for me. First we drove to Hartford, Connecticut, the home of Perkin Elmer, an optical instrumentation company much like Carl Zeiss Jena. Since this was the only American optical company I had heard of, I considered the interview there extremely important; but everything got off to a bad start. I met first with someone who spoke only English. I could understand almost nothing, and he quickly gave up. Then somebody arrived who spoke some German. After reading my resumé and my engineer's diploma, he said several times, "We need you." But we, too, had a hard time communicating. He had brought some drawings with him, and he asked me questions about them. The views and cuts were side-reversed, and all the dimensions were in inches and fractions instead of millimeters and decimals. I became very disoriented. The drawings showed the tolerances for the parts—the largest amount they could vary from the measurement and still be able to fit together. They were given in dimensions from one thousandth of an inch down to 50 millionths of an inch. When he asked me to interpret the tolerances for slip fits, tight fits, and press fits, I got completely lost. It had taken me years to master metric dimensions and the metric tolerance system. I could not grasp immediately this American system. When we left, I knew I had failed. Even his polite "We'll call you" did not make me believe I would ever hear from Perkin Elmer again.

The next day my uncle and I drove to Staten Island to my second and last lead, the S.S. White Corporation, a dental equipment manufacturer that competed with the Ritter Corporation in Rochester. The man who managed the design and development department came down himself. As soon as I said, "Good morning," he asked me if I was German.

I answered in German, "Yes, I just left Germany six days ago." He switched immediately into German and never went back to English. That saved me; at the end of an hour he said they needed fresh blood in his department and I could help him a lot. He said I could start as early as the next day if I wanted to. I took him up on his offer.

Instead of driving back to Long Island with my uncle, I rented a room close to the main road leading to the S.S. White factory, the thoroughfare on which buses traveled from the dock of the Staten Island-Manhattan ferry to the far end of the island. Settling myself in this place solved two transportation problems at once—getting to work and getting into the big city.

All the engineers at S.S. White welcomed me. Although we could communicate only with great difficulty, they were friendly, helpful, and extremely understanding. In Germany no one could have obtained or held a job with such a limited knowledge of the language, but in America one could survive.

My first design project was a high-speed dental handpiece, like those every dentist now uses. A product of this sort had already been invented, but S.S. White wanted its own improved version. I had never designed a turbine before, but I did not care. I needed the job, and I had heard that in the United States anything was possible.

After a few days I settled into a routine—getting up in the morning, taking the bus, struggling with my work at S.S. White, eating my sandwich lunch, taking the bus back to my small room, eating a thirty-five-cent hamburger at the nearby diner or lasagna in the Italian place further down the street for the same price. I learned how to use the many different tricky vending machines, how to catch Coca-Cola in the paper cup from the twenty-five cent automat, and how to do my laundry in coin-operated washing machines and dryers. All this was new to me.

During the first few months I dared not spend money on anything. I lived at the survival level. When I was not at work, I was mostly in my room, which was in an old Victorian house that had moldy, stinking carpets. At least I was warm there while it snowed outside.

I wrote no letters back to Germany. I was somewhat ashamed of what I had gotten myself into—a design job in a low-level factory that appeared to date from the nineteenth century. But I was absolutely determined to accomplish my objectives of learning English and seeing America, and I tried to make the best of everything.

In early March, when it began to be a little warmer, I started to spend weekends in Manhattan. Whenever I took the Staten Island ferry to New York, I felt like a passenger going on a trip around the world. Never routine, this ride across New York Bay close to the Statue of Liberty always inspired me with an energy that would almost sweep me off my feet. The scale of New York itself was overwhelming. At least one hundred times bigger than Jena, it must have been a million times larger than the old inner city, my childhood playground. When I started to walk the streets of Manhattan, I was always in a state of delirium. There were moments when I yelled out loud, "I am in New York, fantastic, can you believe it!"

I would practice my English on anyone. I talked to the passengers on the ferry, to Wall Street brokers having their lunch on the street, and to the beggars on the Bowery. I walked everywhere wearing my black overcoat and my black wide-brimmed Borsalino. I even walked all over Harlem, finding everything fascinating and remaining unaware of any risk. People just stared at me as I did at them. I later found out that I got away with so much apparently because New Yorkers thought I was a priest from a foreign country and therefore untouchable and deserving of respect.

Once some guys did try to lure me into a hallway, but I could not figure out what they wanted and did not follow them. Another time a man invited me into his home after I had talked to him in a restaurant. It turned out that he was gay. I ran out when he mentioned that I had nice "cheeks."

I spent hours and hours visiting the many museums, art galleries, antique stores, and exhibit openings in New York. It became clear to me that coming to the United States had been the right move. It was a great privilege not only to see everything, but to actually live as a New Yorker.

My main problem was social. During this period I made few permanent contacts and no friends. I could not warm up to American women. They seemed quite unlike the women I had loved or found attractive and they seemed to have values very different from mine. They appeared demanding, spoke in high voices, wanted to be wined and dined, did not even kiss you after a nice date, and often wanted to get married right away. Really, though, I think I just was not yet able to adapt to them. Also, I was already thirty-two years old. I wanted to have very young girlfriends, but I was unwilling and unable to spend money on them the way American men did. I tried desperately to find a nice German or European girl, and often went to the German-style cafés and restaurants in Manhattan's Germantown, around East Eighty-sixth Street, but had no luck.

I became lonely and homesick. My thoughts constantly drifted back to Germany and all the women I had loved. I especially regretted not having spent more time with Christa, the girl I had met on the Baltic Sea. Her image appeared more and more often in my mind, and I kept replaying the scene on the beach when I had chased her into the dunes but had not kissed her, and the night we spent together in Leipzig.

I decided to write her. Since I did not know what had become of her, I wrote a letter to her mother. Two weeks later I received a letter from Christa. She also had left East Germany, and she was working as an operating room nurse in Hamburg. Her letter sounded as if she was glad I had thought of her. She wanted to know why I had written her, what I was doing, and whether she could visit me. I got completely trapped in dreams about her. Deprived of any realistic feedback from actually being with Christa, my imagination convinced me that she could satisfy all my wants, cure my problems, and make me happy in this new world. I wrote her about the situation I was in, about my plans to see the world, and told her, yes, she could come, and we could perhaps live together. I knew she would come. I stopped looking at other girls. I had committed myself to this rekindled relationship.

I had several months to wait before Christa's ship would arrive, though, and I used the time to search for opportunities to improve my situation. I was doing very well at S.S. White. I had designed a new high-speed drill: smaller and quieter than the existing ones, it also had a longer life and a turbine assembly that the dentist could change faster. I was promoted. My colleagues were very friendly, except for one man who thought I was competing for his position. With each passing day, he became more irritated until one day he exploded, "Why don't you go back to East Germany?"

Everybody looked at me, wondering how I would handle the situation, but I only said, "Why don't you go? I have already spent enough time there." Everybody laughed.

By now my English had improved so much that I was even trying to read the *New York Times*. One day I skimmed through the employment pages and found an ad offering a job as superintendent in an optical instruments company. I phoned for an appointment and went over for an interview. They liked my resumé and my diploma, and I liked their salary offer of $1,000 a month, which was much more than I was earning at S.S. White. During the interview, however, we seemed to be talking at cross-purposes. I could not understand why they were asking such strange questions. They realized from my questions that I was applying for a different job than the one advertised in the paper. I had mixed up the word *superintendent* with the word *supervisor*: I wanted to run a design and development group, and they wanted someone to take care of all their equipment and their manufacturing facilities. It was very funny because it turned out that they actually did need someone like me. They offered me a job as a project manager. I was so embarrassed, thought, at having made a fool of myself, that I left quickly and didn't take the job.

I decided to concentrate instead on identifying something I could manufacture and sell. I spent a great deal of time in hardware, sports equipment, and department stores looking for existing items I could design better or produce cheaper. I learned what did and did not sell, how much people were willing to pay for various items, and how stores priced their goods, but I did not find a promising project.

Finally the day approached when Christa would arrive. At the very last minute I was able to buy a car, a 1950 hunter green Oldsmobile 88 that one of my colleagues had just traded in for a new one. When he learned of my interest, he snapped it back from the dealer for the $150 trade-in allowance he had been given. He assured me that it was a very reliable car even though it was ten years old and had 100,000 miles on it. It had been garaged all the time, and he had not driven it in winter because of all the salt on the roads. I was very happy to get such a fine car for only $150.

I drove to Manhattan and waited for the ship to come in. This was the moment I had been dreaming about. I waited several hours before I saw the ship, and then several more hours until it finally finished docking and discharged its passengers. At last Christa walked down the gangplank. I was surprised: she looked much older, very pale, and exhausted. She was not the person I had imagined. Not even her personality appeared to be the same.

I believe everyone develops fantasies all the time about other persons, even the ones he lives with. In particular, men fantasize about women, and women about men. Normally, daily contact exposes those fantasies to the test of reality. With Christa living so far away, my mind had been free to compose its own surreal picture. I had lived with my fantasies about her for years, with no corrections or personal contact to force adjustments. When she came down the gangplank, though, I found myself confronting fantasy in conflict with reality. While I was still saying to myself, "What have you done—this is not the person you want," I quickly began saying to Christa, "My God, it is good to see you, how was the trip and where is your baggage and where is the sea kayak I asked you to bring?" I am quite sure similar thoughts crossed Christa's mind because she looked as startled as I did and kept looking back at the ship for help or perhaps to see some nice person she had met on her way over. Meeting each other this way after eight long years was perhaps the biggest surprise of the poorly conceived adventure we were starting together.

Christa was the daughter of a German military officer and was very German. She had grown up in a small town. Our very dissimilar backgrounds led us to see the world quite differently. She once told me that while she was growing up under the tender, loving care of her parents, she had for years not realized that most people had to buy apples, pears, and plums in grocery stores. She had assumed that everyone could pluck fruit for free from trees that grew in gardens like her family's. With her background it must have been very hard for her to comprehend that when I was growing up I had picked fruit out of garbage piles or had stolen it.

It took us several weeks, including a two-week trip to upstate New York through the Adirondacks and into New England, to come to reasonable terms with one another. It helped a lot that Christa was a straight shooter, very candid, very willing to work and take care of me. She found work as a nurse right away, not too far from S.S. White. It was late summer, and we made many weekend trips to the beaches in New Jersey and Long Island. With a suntan, blondish hair lightened by the sun, and her good figure, Christa began to look like her old attractive self. Even so, I remember that once, as I was feeding the blackbirds on the tarred roof right below our window, I felt as if I had lost contact with my own essence and values, and had slipped into living someone else's life.

After three weeks Christa lost her job because she kept washing my laundry at work and hanging it up to dry on a clothesline on the hospital grounds. We moved closer to Manhattan into a small studio apartment with a hot plate and a small refrigerator as a kitchen. Christa got a job in Manhattan. The only guests we had in that apartment were my friend Harry from the office and his wife. Christa tried to fry steaks in a frying pan on the hot plate. They turned out so tough we could not eat them.

I was still ignorant in many ways about American customs. I remember not using any deodorant for the first several months. Eventually some kind person told me that in New York's heat one could not keep friends without using it.

When the winter came, it was one of the coldest I can recall. Three major snowstorms smothered New York, each burying my car so deeply that I could only identify it by the way the antenna was bent. After that winter we decided to take a trip to Florida over the long Easter weekend. To save the car, we drove fifty miles an hour all the way to Florida, stopping only a few hours to see Washington, D.C. We ended up in Vero Beach because we saw our first palm trees there and found a beautiful and affordable little motel right on the beach. We were very happy there and managed to fall in love with each other again.

One morning as we were walking the beach, we came to a pier. From it we looked out over the turquoise-colored water and the white sand beach. Below us we could see many fish and even giant sea turtles. We talked with a man who was fishing off the pier. When I mentioned that I would like to try my luck fishing, he lent me one of his very large rods. He fastened a piece of eel onto its hook, which looked as big as the ones in a meat market.

I practiced throwing it far out into the ocean and reeling it back in. All of a sudden my reel started to spin like crazy. I had no idea what to do. I put on the brake to slow down whatever I had caught, but no matter how hard I tried, this monster kept pulling the line out fast. Since I could hardly hold the rod, the man took over for me. He was a big guy who had one of those belts around his hips with a clamp set in it to hold the rod. He fought the fish for more than an hour, reeling in and then letting line out.

He kept saying, "This is a very big fish. If he gets under the pier, he'll break the line." Fortunately, the fish kept pulling only to one side.

He told us, "We have to bring him in on the shore because he will be impossible to lift onto the pier. Get my big gaff and wait on the beach. If he gets close, hook him."

I took the gaff down to the beach. The fish came closer and closer. I could not see what it was—the big waves stirred up the sand, making the water close to the beach murky. As the fish got closer, sometimes I could see part of it. I tried to hook or spear it, but its skin was too tough. Behind me all the bathers were watching and yelling, as if I were a gladiator fighting a beast. I could not hear what they were yelling at me—the ocean was too loud. I felt like a hero and went farther and farther into the water. Some waves rolled over my head, but I finally managed to thrust the gaff into the fish's mouth. Then I hauled it slowly onto the beach. Once out of its element, the monster lay as immobile as a sunning seal. It was a tiger shark, measured at eleven-and-one-half feet long. The fisherman told us, "In nine years of fishing on this pier, I've never hooked such a monster. It will be the biggest one ever registered at Vero Beach."

The crowd had been yelling, "Get out of the water! It's dangerous! It's a shark!" Not understanding, I had risked my arms, legs, and maybe my life. But it was certainly a lot bigger than the last fish I had caught as a little boy in Jena's Paradise Park.

I asked the man, "What shall we do with it?"

He said nonchalantly, "Oh, nothing. We'll just leave it here. The crabs will eat it."

Christa and I started to walk back the way we had come. Ten minutes later we returned, thinking we should keep some of the shark's teeth as a memento. The crowd had dispersed, and the fisherman was gone. So was the shark. Someone told us that he had hauled the shark away on his truck.

After a few more days in the sun we drove back to New York through Georgia, South Carolina, North Carolina, and Virginia, admiring the peacefulness and grandeur of the great Smoky Mountains, with their endless forests of beech and maple. This trip became a turning point for us. What we saw of the country only whetted our appetites to see more. I had been feeling cramped at S.S. White, partly because of the work I was doing, partly because of the crowded life in New York. So, on the way home, Christa and I worked out a plan to quit our jobs and drive across the country to California. We both wanted to see the Pacific Ocean, ski the Sierra Nevada, and, most important, see for ourselves the giant redwood trees we had once seen on a postcard and could hardly believe.

17

Giving up my foothold in the new world was not easy. When I talked to my boss at S.S. White about leaving, he tried to persuade me to stay, essentially offering me his own job—running the entire research and development department. But working at Carl Zeiss Jena had given me high standards and expectations, and I felt that S. S. White would never be able to offer me opportunities to use all the knowledge and skills I possessed. When my boss confided to me that my newly designed high-speed drill would not even be produced, but instead had been used to negotiate a better deal with the supplier of the existing handpiece, my decision became final.

I resigned and we gave up our apartment. All our belongings fitted comfortably into my Oldsmobile 88. Excited and free of all ties, Christa and I started our drive toward the American West, maintaining our safe speed of fifty miles an hour on the endless roads. It was May of 1961, and the forests and fields looked wonderfully fresh and green. We enjoyed being free of stress and daily routines. We had saved enough money to survive for two months on the road and then to fly back to Germany should we want to. Strangely, I don't remember much of the trip, except the vast wheat fields of Kansas, the snow-topped Rocky Mountains, Dinosaur National Monument in Utah, and the long stretches of road through the desertscapes of Nevada.

I did most of the driving, while Christa, thoroughly exhausted from her job, slept most of the trip, her body slouched against the car door. To kill time on those interminable stretches, I started having conversations with my father. I was now thirty-two years old, three years older than he had been when he died. This was really his kind of life, his kind of journey.

It was not absurd for me to posit that my father was still alive and traveling with me. It was at least as natural as embracing a religion and carrying on an inner dialogue after being indoctrinated in its tenets. Once one accepted the dogmas, one entered a labyrinth of religious thought, a maze of supertitions, suspensions of belief, and contradictions that could leave one lost in one's own mind. My father and I talked about believers' gods and the various concepts of an afterlife. We considered whether there really could be such a future life. We concluded that for those who believe, it exists, at least in this life. I knew, however, that I would never come to that belief; nor would God ever exist for me, because I saw Him as a virtual object created by man himself. Had I lived in earlier times, perhaps before Jesus, it might have been easier for me to become a believer because gods had been more real then. According to the accounts of many religions, people then actually had a chance to encounter a god in person or see him or her in the form of another person. This more sensual, less abstract concept of the divine made more sense to me than what organized religion has made of it.

I also talked to my father about life and death, telling him that I had come to the conclusion that though anything was possible to a living mind (including conversations with one's dead father), after a being's death absolutely nothing remained except matter itself in its original or altered atomic and molecular state. It was incredible to me that the great Egyptian culture had for thousands of years believed in an afterlife partially based on its observations of scarab beetles, and had even built the pyramids because of their belief. In two or three billion years, during which many thousands of life forms had existed through uncountable generations, not a single piece of evidence supported the idea that anything living had ever come back. For me, that set the probability of a return of any life in its existing form at zero. People's belief in or hope for a future life or many lives was to me only a reflection of what great importance humans assign to themselves.

My own theory about life was that humans, like all other living creatures, are links in a chain. Once you have given life to your children and raised them, your job is done and your existence becomes less and less important. For life as such—for the propagation of the species—you are not needed anymore. I have often asked myself whether life has any special importance in the universe. I asked my father whether the state of matter with all its energy stored peacefully in rocks, metals, and minerals was not a better one than the constantly changing state of matter in any life form, including us; but he gave me no answer.

Such conversations and thoughts occupied the three weeks till Christa and I reached Nevada. When we stopped at a roadside café near a little town where we saw real cowboys herding cattle, we ordered steaks. We had seen them advertised on a road sign at one dollar apiece. When the waitress brought the steaks, they truly were bigger than the plate. We were delighted, but regretted that we had ordered two when one would have fed us both and left us one more dollar.

We continued on to Reno that night, wandered through the strange world of the casinos, laughed about the many wedding chapels where one could get married instantly, checked into one of the many inexpensive hotels, and slept well.

The next morning we drove on toward San Francisco. Just as we saw the sign "Welcome to California," we heard on the radio that the Communists had cut off free access to East Berlin and were building a wall along its border with West Berlin. We were stunned. What I had anticipated in prison had come true just three years later. With this event our former world had been cut in two. All we had left there—family, friends, memories, and the hope of going back sometime—had been brutally cut off. I knew at once that from now on the cold war would become much more serious, and the third world war constantly predicted by the eastern regimes much more of a possibility. We were glad to be in the West, close to the Pacific Ocean, because this seemed like the last place the Bolsheviks could reach, and safer by far than West Germany.

When we got to Sacramento, we visited the state capitol building, but since we were tired of sightseeing, we decided to cut our tour short and drive straight on to San Francisco. We spent only a few hours in the city, mainly walking the streets around Union Square. We ate dinner and drove around a while to see other parts of the city, but I had the feeling I would never find a job there. San Francisco had no industry, and everything was so expensive that we knew we could not survive there long with the little money we had left. I studied a map to see how to get out of town to find an inexpensive hotel. When I noticed the name "Redwood City" about thirty miles from San Francisco, I imagined it in the middle of a majestic redwood forest. I said to Christa, "This is it, this is what we came for. Let's drive there, and tomorrow we'll wake up and walk among the big trees I showed you on the postcard."

We drove south on Highway 101. It was eleven-thirty at night, and pitch dark. We were tired. When we saw the sign 'Redwood City." we turned off and drove onto El Camino Real. We saw no trees, but in San Carlos, the next town to the north, a small hotel advertised "pool and room for two, twelve dollars a night." We spent our first night in California there.

After breakfast we drove up and down El Camino and onto side streets looking for redwoods. There were none: they had all been chopped down in the late 1800s. A sign told us, "Redwood City, Climate Best by Government Test." We decided to stay and look for jobs. I picked up a paper, saw a little employment ad for an engineer, went to the company, got the job, and started the next day.

Optics Technology, located right in San Carlos, was a start-up company designing instruments that used fiber optics—a completely new material I had never heard of. The company's founder, Narinder Kapany, was a Sikh; he had a long beard and hair that had never been cut stuffed under a huge turban. He gave me my first assignment—design a laser. I had never heard of a laser, either. They had just been invented. Since several physicists worked for the company, I asked Kapany whether I could work together with them to develop the laser. He told me, "No, I want you to build one without their help!"

This assignment scared me. I wasn't sure that I alone could reinvent a device that someone else had just invented. I was no physicist, and I didn't know whether I would be given enough time to learn the underlying physics before having to produce a laser design. I was afraid to get into a project at which I could fail miserably. Kapany seemed to be a kind man, and I felt that I could ask whether there was any other project for me to work on. Kapany showed me the idea for a sugar refractometer based on fiber optics. I knew how such instruments worked; we had built them at Carl Zeiss Jena, but I thought the ideas and principles Kapany showed me would never work, and I told him so. He said, "Try it anyway," but I had lost my faith in him.

At lunch time on my second day I drove along El Camino south to Palo Alto, searching for another place to work, and accidentally took a turn onto Page Mill Road. I wondered who occupied the beautiful new buildings there, so I drove into one of the driveways, which happened to lead to Hewlett-Packard's Building One. I had no idea what Hewlett-Packard was or what it produced, but I was approaching my job hunt as I had done when I walked across Germany with my backpack, knocking on doors. I just parked my car and told the receptionist I was looking for a job. She called the personnel manager and he and an engineer interviewed me. When he offered me a job, I accepted. I then returned to Optics Technology and quit.

Meanwhile Christa had found a job as a nurse in a Redwood City hospital. That same day we rented a small apartment close by so she could walk to work.

I never felt proud of having left Optics Technology the way I did. I should have given Kapany and myself more of a chance. Looking back on it, I probably could have made a difference there, designed a beautiful laser, received some stock options, learned how a small start-up company works, and perhaps been prepared by the experience to set up my own start-up company early on. At the least, I would have made money on the options.

What happened instead was that I learned a great deal at Hewlett-Packard from its high industrial ethics, its very high standards in developing and producing products, and the management skills of the people I worked with. By letting fate lead me to a job at Hewlett-Packard, I had lucked into a terrific situation. I always considered Hewlett-Packard to be the electronics equivalent of Carl Zeiss.

The man who hired me, Ed Phillips, set a new standard for me. He was the best engineer I had ever met. He had served in the army in Germany after the war; he was one of the guys who brought home a German bride. They had bought a big house, furniture, and cars without a down payment, and were raising three kids. Extremely intelligent and generous, he was very American. When he decided to take a vacation in Germany to revisit the area where he had been a tank commander, he figured out how much the trip would cost him per hour by combining all travel expenses with earnings lost during the vacation time, and then dividing by the number of hours spent abroad. No German, I thought, had ever done this. Germans took six weeks of vacation every year and did not live to work, as the Americans seemed to, but worked only to live and pay for life.

In some ways, Ed was like me—very inventive, always searching for ideas for starting a new venture and making money. He was also somewhat of an eccentric, the only engineer at Hewlett-Packard who ran around wearing a sun visor, like somebody at a racetrack. He had a hard time setting limits for his imagination and inventiveness. If you had asked Ed to design a house, he might have come up with a beautiful one which, unfortunately, would have cost you far more money than you could afford.

When we worked together, I think I often was better able to assess a project's probability of succeeding, and to project the amounts of effort, time, and money required to pull it off. I also made decisions more carefully. But unlike Ed, I was not really a good designer on paper; I was more of an inventor and manufacturing engineer.

In those days I often had difficulties working with inch-based American dimensions and tolerances; I had to convert measurements into the metric system. This led to a near-washout for me in my first four weeks at Hewlett-Packard, but my ideas were good, and that saved me from being fired.

I also had to alter my German style of telling people what to do. When a machinist told me he could not fabricate what I had designed, I asked him to let me use his machine to show him. I made the part myself. I almost got fired for that as well. I did not realize that in America one had to ask someone politely, and only suggest how to do a job. Unlike the Germans back home, who were used to being told what to do, and when and how to do it, Americans wanted to make decisions themselves and have more elbow room. In fact, I could hardly believe the great latitude employees were allowed at Hewlett-Packard. Engineers could fail without falling into disgrace or getting fired, as would have happened in Europe. I repeatedly witnessed project managers spending a hundred thousand or even a million dollars on projects they failed to pull off. They usually got a new assignment, and, if they failed again, maybe another one as well. Success was judged here by the quality of the effort and performance as well as the actual outcome. You could fail and still be a hero, as long as there was a very good reason for the failure.

Hewlett-Packard had a unique, modern management style. Most surprising to me were the attitude and demeanor of William Hewlett and David Packard themselves. Their company was medium-sized at that time, with the two founders already worth hundreds of millions of dollars, but Mr. Hewlett and Mr. Packard were addressed as "Bill" and "Dave." They were seen, and their presence was felt, every day. Their influence and advice entered all decision-making and every project. One never had the feeling that they were talking down to you. They had a "we are all in this together" attitude. I can still hear them saying, "We are here to do great things, only the best." And, "If it's difficult, we want to do it!"

I met them several times in the cafeteria as they stood in line to buy lunch like everybody else. On occasion I would sit with them and discuss projects right at the lunch table. Amazingly, nobody ever questioned whether, or acted as if, I might be a spy. For me, not long out of East Germany, this was surreal.

Once we were working on a very complicated set-up for a microwave spectrometer when Dave Packard came down to our department to see how we were doing. All of us were absolutely flabbergasted when he stepped up to the set-up and started twisting knobs, operating it as if he had been working on it for weeks like us.

I did have one small run-in with Bill Hewlett. One Sunday when I had come in to the lab to work on my project, I went out to get a sandwich. Bill, who normally worked over the weekend on his big ranch in Gilroy, walked in while I was gone and was sitting at my workbench repairing his radio when I returned. He had ruined my entire set-up. I complained, and Bill apologized. But to me this was America, and I loved it. Here was multimillionaire Bill Hewlett in his farm overalls, worrying about his tractor radio and repairing it himself, even though he had hundreds of electronic engineers on his payroll. Attitudes such as this made up the California spirit. Nothing of the kind existed in Europe, nor would such attitudes have been allowed to take root—particularly not in the areas controlled by the Communists. If I had ever returned to East Germany and told these stories, they would have hanged me. Back there the government condemned capitalism every day. It was busy freezing initiative, stifling thought, and killing the country.

With my family still in East Germany, I kept up on the state of affairs there. We managed to maintain contact, although the only communication permitted was by letter. The East German authorities opened mail, some of which disappeared undelivered. Neither my letters nor theirs could contain anything political, any negative comments about the East German government, or any description of living conditions in East Germany. I was, however, able to send them food packages and other necessities. I remember at one point sending "Wedding Gift" packages containing everything newlyweds need, such as bed sheets and towels, which they had been unable to obtain.

Getting money to my family was much more difficult.

The process showed what odd conditions could be produced when a government tried to completely control its country's currency and economy. My family could not legally receive any western or eastern currency without going through official channels. It was tempting to buy East German currency in the west, where one could get ten Deutsch Marks East for one Deutsch Mark West, but it was too dangerous to enclose the money in a letter to my family. I could, however, get them Deutsch Marks East if I bought them from the East German government at the government-set exchange rate of one Deutsch Mark West to one Deutsch Mark East. Setting an equal exchange rate was meant to show that the East German government's money was every bit as good as that of West Germany. It also allowed the East German state to pocket 90 percent of the value of any western money exchanged.

We finally found a better solution. I bought my brother and mother automobiles. In East Germany, an ordinary citizen who wanted to buy an automobile had to pay for it and then wait about twenty years before he got his automobile. However, because the East German government needed foreign currency badly, it allowed individuals on the outside who could pay in a hard currency to bypass the waiting list and get an East German car delivered to relatives immediately. In an economy of government-created scarcities, my brother was able to sell a car I purchased for him and my mother for about twice as many Deutsch Marks East as a person would have paid to be put on the waiting list for a car. In this way, I was able to provide for my mother and help my brother. I disliked helping my former persecutors by giving currency to the East German government, but my feelings were less important than my family's needs.

At Hewlett-Packard I met many other European expatriates, quite a few of whom were more academically educated than I. They came from Europe's best universities and were usually the brightest graduates, or H-P would not have hired them. I liked many of them and started a sort of European-American club to enable us to exchange ideas and viewpoints. I would have liked to have added some writers, poets, or painters, but attracted only German and Swiss engineers and scientists, about twenty-five in all. Most of us were new to this country and glad to hold on to one another.

Our discussions were mostly philosophical, but they centered on a very finite question— whether to stay in America or go back to Europe. Adaptation to American standards and values seemed out of the question to us. Our European upbringing, values, understanding, and culture could not be abandoned or traded in for the American equivalent. We complained about how superficial Americans were: they could not follow European thinking or grasp our aesthetic values. While the Americans loved it when we talked to them about Europe, we knew it was like handing somebody a photograph of a beautiful beach: they could never feel the sun, smell the water, or hear the sound of waves washing over sand and pebbles. Americans had few conversational skills, sang different songs, and had learned different poems than we had. They had peculiar senses of humor, and their jokes were not as funny as ours. They liked baseball, a boring game, and football, a rough game, while we loved our beautiful, exciting soccer.

Sometimes we complained about the language, which said many things so much more awkwardly than our own. For years I fought with the different meanings of English words and idioms. I could never bring myself to say "to make love" or "making love." I was still translating a lot of phrases literally into German, and in German "making love" sounded as if one was fabricating something, or acting mechanically, without mental involvement. I hated the idea of what I took to be the separation of love and sex.

We all had trouble recognizing and admitting that America had developed, in many ways, a more advanced civilization than our European one. At the same time, none of us could deny that America had become the leading nation in the world, the greatest democracy, the possessor of the greatest freedom of the press, the leader in music and entertainment and in high technology, too, and that the rest of the world was copying America's lifestyle, leisure fashions, fast food culture, and even its bathrooms.

The truth was, we all wanted to have our cake and eat it, too. We all liked both America and Europe, and we were each in the midst of a period of personal adjustment. As time passed, our behavior, thoughts, and attitudes became more and more Americanized. Most of us eventually changed our orientation, became good Americans, and contributed to the rabbit-like multiplication of new companies and new technology in California. Only a few went back home to Europe. Four years later, the club dissolved after we had discussed all our subjects to death, and soon after some members had begun bringing women to the meetings.

I think now that any immigrant coming to America needs a five-year investment period to make American values his own, to learn the language, to understand the rules of American business, to appreciate the qualities of American women, to make some real friends, to know the country, to feel at home, and to start looking for the opportunities to do more than just hold a job and adhere to prescribed professional routines.

Acculturation is, however, never really a thorough process. While every immigrant knows the whole library of life of his own country, he can never read all the books in the library of the new continent. An immigrant has to infer what is natural to the American-born, what Americans mean when they talk, and what their experiences have made them. In addition, an immigrant has to struggle hard to catch up to American standards of living. There is no easy way, and the job is never complete. Still, after some time, the immigrant quite typically falls in love with America and reaches the point of enjoying even those very peculiar American institutions which could only be exported to another culture if you supplied a lecture and footnote to explain every joke—for example, a Woody Allen movie.

Christa and I spent only the first six months of our Americanization process in Redwood City. Driving home from work late one night Christa almost knocked down one of the posts supporting our apartment building. It shook, causing a panic among the tenants. I had been sleeping and it felt as if there had been an earthquake. After that we moved to Palo Alto. There, one night Christa drove the car clear through the back wall of the carport beneath us. She had other accidents with my car, and I finally sold it. Amazingly, after 50,000 more miles and with serious damage, it brought $25 more than I had paid for it. Maybe never going over 50 miles an hour had paid off.

The apartment house in Palo Alto had a small swimming pool. We enjoyed that tremendously.

On a ski trip to Reno, Christa and I got married. This was more a cosmetic measure than a serious decision: it legitimized our situation, making Christa more acceptable to our married friends—a foolish reason for entering marriage. I had no plans for a family, liking at that time neither kids nor family responsibilities. We sometimes had arguments about small issues such as her smoking and my habit of driving the car on empty to the very last drop. Still under the influence of my rough prison life, I was quite insensitive to her feelings. In my ambitious search for treasure, I was roughly bulldozing everything in my way. This loaded our relationship with problems, but neither of us realized it. I never felt completely settled in this relationship. One day while I was buying groceries, I met a very attractive young woman. I found her fascinating and actually wanted her. I never saw her again. But this woman had developed enough interest to call me at home. I was not there, but Christa was. The woman asked Christa whether she would mind if I went out with her. Christa was furious, but she only told me about this incident much later.

In a lonely moment in our relationship I sent a postcard to the girl I had met at the United States Consulate in Frankfurt. She replied with a long letter. Christa intercepted it and hid it from me. I did not realize how these small actions and episodes could put hairline cracks into the fragile porcelain of a marriage, or how, once the structure was cracked, it could shatter into pieces when pressure was applied.

During this period I became very interested in a book about Schopenhauer's life and philosophy that my friend Ushi had given me shortly before I left Germany. While Christa worked night shifts as a scrub nurse in the Redwood City Hospital or was busy cleaning, doing laundry, cooking, and running errands, I tried to extract Schopenhauer's ideas from page-long German sentences. What fascinated me most was how precisely Schopenhauer had laid down his thoughts, developing a unifying philosophy that included every aspect of life and spirit, physics and metaphysics, art and wisdom—and all by the time he was thirty.

This book gave me a lot to think about, including the fact that I was past thirty and had accomplished absolutely nothing. My studies came to an end one day when we were expecting friends to visit. I had left my book on the coffee table. Christa accused me of displaying it there to impress them.

One Sunday as we drove through Los Altos Hills, looking at its expensive homes, I said to Christa, "Someday I want to have a house like this."

When she replied, "You will never make it so far," our relationship developed another crack.

Shortly thereafter, Hewlett-Packard decided to shelve my beautifully invented, designed, and executed line of very high frequency microwave equipment because the market was not yet ready for it. I went into a twilight zone of uncertainty. I was frustrated. None of my designs had ever made it into production. Other people always made the product development decisions and told me what to do. I promised myself I would learn more about marketing to become more capable of predicting and understanding their product decisions.

I had driven myself hard and was very tired. The factory walls were moving in on me. I knew I had to give my Bohemian side some air. I had saved some money, and I decided I could invest half of it in the stock market and use the rest to take a trip. By buying on fifty-percent margin, I hoped to find when I returned that I had more money than when I left. I wanted to see Germany again, and I remembered my resolution in prison to see the entire world to compensate for the time and the living I had missed by being confined there. A trip around the world sounded like the ultimate compensation.

I talked Christa into letting me go on this adventure. We did not consider having her come along. Since I did not have very much money to spend, I would have to travel without concern for comfort. I planned to go through treacherous territory, to places I knew only from reading Karl May. I would travel off roads, where tourists could not be found. No threat of hardship or harm would scare me away from a region I wished to visit. I was willing to risk my life for this experience. I would go through Europe, the Balkans, the Middle East, Africa, Persia, Afghanistan, India, Thailand, and Japan, and then return to San Francisco.

I made big posters advertising for somebody brave enough and tough enough to travel with me and placed them on the billboards at Stanford and at the University of California at Berkeley. Nobody replied. People probably thought I was either crazy or not serious. I would have to go it alone.

18

In March 1963, Hewlett-Packard gave me a six-month leave of absence for my trip. I said goodbye to my friends; after hearing my plans, they worried that I might not make it back. One guy, who had always been very friendly and had always had his eye on Christa, said, "I hope you don't come back!" He apologized right away, but I thought that was a clumsy attempt to cover up his Freudian slip. I wondered how he could think that way; to me other men's wives had always been taboo.

Christa brought me to the airport. We hugged and kissed for a long moment, and she cried. I felt bad about leaving her behind, knowing I could have chosen an easier itinerary, or just traveled to Germany with her; but this trip was part of my prison program, and I felt I needed it to get myself breathing again. Then the farewell was over, and I was seated by the window on my way to Portugal. It was good to know I could return anytime and have a woman as solid as Christa waiting for me in California.

As I flew over my new homeland for the first time, I looked down on the route I had driven with Christa. I was now glad to be part of this country, which appeared incredibly big, beautiful, and indomitable from up here. I felt lucky to be setting out without any particular plan, reservations, or time limitations. I could do everything totally spontaneously, finding my path as I traveled. At the end I hoped to meet Christa for a delayed honeymoon in Hawaii.

I enjoyed two days in Lisbon, seeing the sights, eating the fine food, drinking the excellent wine, and being served by superbly trained waiters. I was finally living as I had always wanted to, completely free, with an attitude like Thomas Mann's Felix Krull: the world was mine to enjoy.

There were limits, however: I had only about $2,000 with me to keep myself fed for six or eight months and move myself around the world. With all the sights of the world to feed me, though, I would only need simple transportation and safe places to sleep.

I traveled by bus and train to Spain—to some lost places like Badajos, and to all the famous cities: Seville, Cordoba, Granada, Toledo, Madrid, and Barcelona. What struck me most was that in the little villages and even in the cities the inhabitants went into the streets at night to stroll, stand in groups talking, or sit in the many cafés and restaurants socializing with their friends and neighbors. Because mass tourism was not yet common and I had come in the off season, I was often the only tourist in town. I had conversations with many people, mostly using sign language. Everybody seemed to have enough time for me, with everything being *mas o menos* ("more or less"). This life was the antidote I needed for my hard-driving life in California. Even though I was just a foreigner passing through, I had more human contact here than back home in Palo Alto. I wasn't able to slow down to match Spain's tempo, however; I had trouble adjusting to the fact that trains or buses did not run as scheduled, and I often raced from sight to sight, from the bullfight to the Prado, from buses to trains, as if I were still on my old schedule. Not even the constant rain slowed me down, so I often got completely soaked.

I had started to write a journal for Christa, but I left the first thirty pages on a bus. Then, on the train to Paris, I noticed my camera was missing. I realized I had left it at a restaurant close to the railway station in Barcelona. I could not afford a new camera, and I could not bear the thought of going around the world without taking pictures. From Bordeaux I took a train back to that restaurant. I left my suitcase with some lawyers on the train, who promised to drop it off at my Paris hotel. The restaurant owner handed the camera back to me, somewhat reluctantly. I think he had intended to adopt the orphan. I was relieved, but with this detour the trip from Barcelona to Paris took thirty-three hours.

After traveling inexpensively for three weeks in Spain, I was tired when I reached Paris. At first I found the French and their capital anything but charming. I was already saturated with impressions of Spain, and concerned about how expensive Paris was. It was still raining, and the Eiffel Tower looked to me like a somewhat outdated gizmo.

I was also upset because I didn't have my suitcase. The lawyers had left a note at my hotel, telling me the name of their hotel, but I could not decipher that name. I lived without my suitcase for two days, worrying about it, until finally they called once more and I was able to retrieve it.

Now I went absolutely crazy visiting the Louvre and Paris's many other museums. The incredibly beautiful paintings of the impressionists and Van Gogh, and the vibrant works of the great modern painters inspired me; they reinforced the wish I had formed in New York to devote part of my future to painting, to trying somehow to get my soul on canvas.

From Paris I traveled to see my old friends in Mainz, who lovingly brought me back to life, even washing and ironing my laundry before sending me on my way. It felt good to be back in Germany. In Frankfurt I picked up the brand new Volkswagen I had ordered and paid for from California: I planned that it would provide me with inexpensive transportation and shelter.

When I had last visited Frankfurt, in 1948, everything had been in ruins. One could ride for ten minutes through the center of town without seeing a single house still standing. The population then seemed to be made up of people cleaning up the rubble, black marketers, returning POWs, and refugees from East Germany. Now I found the town completely rebuilt, with fine stores lining the streets, everything impeccably clean and orderly, and all the people dressed neatly and tastefully. I was so impressed that I decided I would stay only two more years in the United States and then come back home to Germany. I was still having difficulty adjusting to American life, and I had not made much money.

I then received a letter from Christa. My stocks had taken a dive, and, worse, her father in East Germany had suddenly died. I called her. I told her that she should try to get a visa and visit her mother on her own because I could not go back to East Germany. My mind filled with concern. There was no way to predict how long it would be before the East German authorities decided whether to give her a visa. If she decided to join me on my trip after visiting her mother, I would have to shorten it and leave out the most dangerous parts.

As we talked, I was also worrying about the high cost of a phone call from Germany to California; our conversation was short. International phone calls were not allowed for in my tight budget. Even paying for Christa's trip to Germany might leave me somewhere on my path without enough money to get home.

I think I was too greedy to give up any part of my trip, and the whole problem seemed too complicated to solve on the telephone. To this day I cannot forgive myself for not choosing the simplest and most appropriate solution—cutting short my trip and flying home to give Christa support.

Instead, I began an inoculation program against all the possible diseases and infections I might encounter. When one doctor told me that just to be reasonably well prepared, I would need at least thirty-four shots over a period of four months, I gave up. I did not have the time or money for such a regimen, and I feared that all those shots would leave me more dead than alive.

My Volkswagen improved my life dramatically. I felt I now had the power and freedom to drive anywhere and to do anything. I bought extra snow tires for driving through desert sands and high mountain snow, and extra-large water and gas tanks to make me less dependent on water holes and gas stations. When Christa decided not to come with me, I sold the passenger seat. The back of the rear seat just happened to fit into its place next to me and made, together with the back area, a wonderful one-man bed. I ended up sleeping 120 consecutive nights in my Volkswagen. Each night I would just park in front of the finest hotel or the railway station, put sheets of cardboard in all the windows, and sleep safely under the watchful eyes of the hotel staff or even the police. Once, though, somewhere in Italy, I almost killed myself when I parked on a bridge, not noticing that railroad tracks ran directly beneath it. I woke up in the middle of the night, dizzy and gasping, with a gigantic headache. I was completely engulfed in the thick black smoke rising from a locomotive parked right below me.

After several weeks of sightseeing in Austria and Italy, I took an auto ferry through the Strait of Corinth to Athens, where I spent countless hours admiring the ancient Greek and Roman sights I had always wanted to see. One evening, all dressed up and walking the streets of Athens, I noticed that some important event was going on in the French Embassy. I walked right in past all the security guards. Nobody stopped me, apparently because they thought I was from the secret service. Soon I found myself in the main hall where the festivity was being held. I had no idea what I was doing, but in this naive way I shared the air with Charles de Gaulle and the King and Queen of Greece. I left as I had come, without incident.

From Athens I went to Istanbul, where at three o'clock in the morning some guys began trying to break into my car. When I pointed a tear gas pistol at them, they ran away. I had no idea how dangerous this was. If they had had real guns themselves, they might have shot me.

As June ended, I was traveling through Syria, Jordan, and Lebanon. These countries were absolutely magic. I loved the people and the way they dressed, lived, and talked. Life was real here, down to earth, natural. I talked to men out in the fields, in the villages, and in the cities. Our communication was mostly with sign language or a few words in German, English, or French, but it worked very well. People everywhere invited me for tea and little things to eat, mostly cookies or melon. I became one of them, sitting in cafés or just on a sack of beans in the stores, smoking the water pipe. Each time they realized I was German, they became enthusiastic. Sometimes this was very embarrassing to me—often they celebrated me as a Nazi, asking whether I had helped to kill Jews or wanting me to read out loud from Hitler's *Mein Kampf,* which some of them kept in their houses as we might keep the *Encyclopedia Britannica.* One man took me with him to his mosque for prayer. I knelt down and bent my head to the ground like the others.

Near Baalbek in Lebanon, I wanted to photograph some nomads I saw, and made the mistake of walking into the desert toward their tent. Three big dogs ran at me and pinned me down. It was noontime and extremely hot. Each time I tried to move, the dogs would bare their vicious teeth and move a little closer. The nomads did not call their dogs back for hours; my knees were shaking, and my fair skin was burning to a crisp. Finally they whistled, and the dogs let me go back to my car.

In beautiful, guest-friendly Lebanon, I met a family with a very beautiful seventeen-year-old daughter, who fell in love with me. For two days she hand-fed me the most delicious foods and made me feel incredibly comfortable and satisfied to be with her in these very romantic surroundings. It was all platonic, of course, but I was really sorry to leave her behind. She could have introduced me to a brand-new life in Lebanon. I began to realize that I was like a chameleon. I could blend in anywhere to live with and like any of all these different people I met, sharing their customs and food. Meeting so many people in cities, villages, and fields in so many countries, drinking and eating what they did, observing their lives and trying to live with them for a while, I was able to see and touch a different kind of being human than I had known in Germany or the United States.

I had seen so many wonderful things by this time that I felt the trip had already rewarded me richly. I had been especially impressed by Rome, Athens, and Istanbul. I could not believe how much in each had survived from the distant past. In fact, all the ancient cities I visited on my way offered living lessons in history, teaching through all five senses lessons on how people lived their lives two, three, four, even five thousand years ago. It was a course of study far superior to anything one could get from books or a history teacher.

I began to regret not having brought Christa along. As I traveled, I thought about improving myself, making myself into a better man who could please her more. I missed her and could not wait to reach towns where a letter might be waiting for me. I filled more and more journal pages and kept sending them off to California. I had other concerns as well. I badly needed a *carnet de passage* to enter India with my car, and I was spending my money quicker than anticipated. I inquired at each town, but there were seldom letters from Christa waiting for me. I could imagine that having had to handle her father's death without support had put her into a tailspin. But I could not help her where I was, and I was determined to go through with this adventure.

In Jordan I tried to photograph some nomads who were cutting wheat they had sown in between the rocks. It was a beautiful sight. One could observe the whole process of harvesting in the traditional manner: cutting wheat with a sickle, threshing it out, and separating the straw from the grain. But as soon as I approached with my camera, an older man came up to me, pulled his curved dagger, and pointed it at my throat. I felt the steel piercing my skin and blood running down my neck. I thought this was going to be my last moment. I stared into his fiery eyes and stayed as calm as I could. He demanded my shoes, slipped into them, walked back to his family, and continued throwing wheat up into the air to free it from the chaff. It took me two hours to bargain back my shoes. He did not want to give them up, but I needed them—they were the only ones I had. But after all, the experience of walking barefoot among the rocks, worried to death about my life and shoes while taking pictures of this beautiful, colorful family doing their age-old job, was a great one.

I will never forget Jerusalem or Bethlehem. With almost no foreigners there, life in these old cities seemed much as it must have been two thousand years ago. If I ever came close to believing in the Christian idea of God and His divine son, it was in the old village of Jerusalem, among the weathered buildings, covered passageways, narrow paths, and stairwells. I found myself reacting as a devout Christian might be expected to react: Jesus seemed somehow still alive here. He became as human for me as the men who were walking in the streets with their donkeys beside me.

I traveled with images of these holy places for days. But they became partially obscured when I saw and climbed the pyramids, and more so when I stood before the big temples at Luxor and Karnak and the tombs in the Valley of the Kings. Confronting the sheer size of what people had been inspired to build was like a powerful blow, shaking up my almost atheistic way of thinking. In the whole universe, I thought, one could never find another place like Egypt. Here human respect for superior powers and the deep desire to be reborn had found its ultimate expression. Their monuments showed that the human imagination could conceive projects on the largest scale. But viewed through the lens of modern rationalistic thought, these monuments also show how limited and ineffectual we humans will always be in our attempts to respond to questions and fears about the infinite with finite creations.

The train trip from Cairo to Luxor and back gave me a marvelous view of the Egyptian way of life—the fields and gardens, the many palmscapes, and the Nile. But it felt quite a bit like standing in a wind-tunnel simulation of a sandstorm: because of the heat, we had to keep the train windows open even though the locomotive often stirred up large clouds of fine sand and dust and then pulled the train through them. After many hours on the train, I looked like the desert itself. I was very grateful to a friend I had met in Cairo for his gift of a large water jug made of unglazed clay. From then on, I always traveled with this vessel, which kept the water wonderfully cool by the evaporation on its outside surface.

In Luxor I lived through an incredible sandstorm. For about an hour, one could see no more than two or three feet ahead as the grit-filled wind picked up everything in its way. Before the storm, I had been unable to walk on the shore of the Nile because everybody seemed to go there in the morning to take care of his sanitary needs. No one seemed bothered that during the day a big market selling everything imaginable, including all kinds of foods, operated right next to this public toilet, or that after the storm, everything had blown over everything else. But I loved everything I saw of Egypt—the Nile, the sand, the palm-filled landscapes, the nice people, and the history. It gave me a great high.

I began to calm down after driving two days farther along my route through the boiling-hot desert toward Baghdad. I found the desolation and peace of the desert just as grand as all the wonderful sights and people I had just seen in Egypt. I will never forget this two-hundred-mile stretch of road with no gasoline stations and only two water holes. In two days I saw only three men. Two were sitting on the road in the middle of a huge area of desert entirely covered with pitch-black stones. When I stopped at a water hole, I noticed the third in the distance, riding a camel. As he rode toward me, it was exactly what I had seen in the movie *Lawrence of Arabia*, only this man was not a Tuareg in black, but a Bedouin in white. When he came close, I stared at his unbelievably brown and wrinkled face and very dark eyes. When he reached me, he said, "*Salaam*," made his camel go to its knees, climbed off, and led the animal to the partially algae-covered watering hole. The camel stepped into the murky water with its front feet and started to drink and piss at the same time. After his camel had finished drinking, the man himself bent to the water and drank from the same spot the camel had used, just pushing the surface algae aside with his hand. I felt like offering him some of my clean water, but he had already climbed back on his animal. He said "*Salaam*," and rode back in the same direction from which he had come. I watched him disappear in the overheated, scintillating air that in the distance looked like layers of water. I knew from my education in optics that this was an illusion, a mirage, but it was still magical. Then there was silence once again, dead silence. It was one of those rare and privileged moments when one can hear nothing, and I loved it.

Baghdad turned out to be the city of a Thousand and One Nights, with more of the fabled Orient of the Middle Ages visible than anywhere I'd seen before—even in the best movies. Everything was real and unaffected by Western culture, especially in the quarters close to the great mosque. All the women wore black robes and were deeply veiled. Some men, apparently from the country, rode their horses right into the restaurants. The hustle and bustle everywhere never ceased. When I walked into the Golden Mosque, one of the holiest places of Islam, guards recognized me as a Christian. They jumped me, beat on me, and dragged me through the dust to the outside. By crawling very fast on my hands and knees, I escaped the agitated, loudly yelling crowd. I was so disturbed and frightened that I started at once to drive toward Tehran.

Getting there turned out to be harder than I'd expected. Thirty kilometers outside Baghdad several tanks blocked the road. In the distance I could see airplanes bombing the Kurds. The soldiers would not let me pass. I decided to drive back a little bit and then cut off the road and drive like a madman in a twenty-mile arc through the desert and back onto the Tehran road. I was so afraid the airplanes would attack me that I drove full throttle, as fast as the Volkswagen would go. After I finally slowed down a bit, I noticed a jeep approaching from behind, carrying two men in underwear. They turned out to be the border guards who had been sleeping on the job. They made me drive the twenty-five miles back to the border to get properly processed. We drank some bottles of soda together, and they let me go on.

It was now August, with the desert temperature always above 110 degrees Fahrenheit and often climbing to 135. I could drink twenty bottles of soda a day and all the water I could get down, and sometimes eat twenty ice creams as well, without ever going to the bathroom. All moisture went out through the skin. The air was so dry one could jump into water fully dressed and be dry five minutes after emerging. The steering wheel got so hot I could not touch it until I wrapped it with cloth.

After driving another seventy miles through a forbidding landscape, I saw a building beside the road. I stopped and went in. It was full of tribesmen with dark, sharp-featured faces, colorful clothes, and turbans with fringe hanging down. I found I was still in Kurdish country, in a place so backward that some of the men had big clubs with spikes on them, like medieval weapons. I was thoroughly scared but made a hand sign for drinking water. A man went out with a tea can and brought some water from the creek. I waited ten minutes and left without drinking it. About twenty minutes down the road, I got stopped by two men who looked like the ones in the inn. They had very old rifles, which they pointed at me. I picked up two apples, opened the door, and gave one to each man. They smiled, looked through the car, got in, and waved me to continue down the road. They got out of the car shortly before Hamadan and gestured me to drive on.

I was relieved when I finally reached Hamadan, but this was not my day. As I was driving though the village, a ten-year-old boy, coming from a bakery with a pile of flat breads in his arms, crossed the street and ran right into the side of my car. In five seconds I was surrounded by a yelling, gesticulating mob, some of whom tried to pull me out of my car. Fortunately, the police rescued me. I spent three days under protective custody in prison, eating delicious food and playing cards with the guards. On the fourth day a judge sentenced me to a negotiated settlement of ten dollars. The judge explained in perfect German that even though the accident was not my fault and the boy had not been hurt, he had to find me guilty; otherwise, he could not live here anymore, since most people in the village were illiterate and did not understand the law.

I drove on to Tehran. Since leaving Istanbul two months ago, I had not received any letters from Christa. I was really worried. I waited in Tehran for a few days; still nothing arrived. I drove on to the Caspian Sea. One day I parked at a very lonely beach and fell asleep by the shore. When I opened my eyes, very large birds were circling over me and sometimes buzzing just meters over my face. They were giant eagles. When I sat up, I found myself surrounded by a large group of vultures. They were probably waiting for me to die. I was not scared because these birds would not hurt anything showing signs of life. I spent hours on this beach watching the eagles soar. I thought it surprising that humans had not invented airplanes sooner. Even before the invention of an engine, they could have observed the best birds soaring, measured and weighed them, and built scale models of bamboo and papyrus. Imitating natural flight would have been much easier than inventing the wheel: nothing like a wheel existed in nature. Since it had taken man so long to invent something as obvious as an airplane, I thought there had to be plenty of other similarly useful devices man had not imagined and constructed. I promised myself to work even harder to identify these when I returned to the States.

I continued along the Caspian Sea all the way up to the Russian border and then eighty miles through very dry desert toward the holy city of Meshed. People told me there were areas in this region where it had not rained for seven years. I saw many camel caravans and camel stations on this stretch of road, which was covered with fist-size split rocks. By this time the fenders of my once-spotless Volkswagen sedan, hammered by the rocks, were completely shot, so full of holes and tears that I could see the coarse profile of my snow tires right through them.

I loved Meshed but could not linger. Once again no

letter was waiting, and I was running out of money.

My keenest dreams of adventure became reality in Afghanistan. I immediately loved its proud and friendly people and the tiny shops on the village roads, where everything was manufactured exactly as it had been for hundreds or even thousands of years. As in the Middle East, I was amazed by how unchanged most of the land was, even though people had been living there for tens of thousands of years. It made me think about my work: as an engineer I was part of the forces working to change the earth. What would people from industrialized areas do to alter this landscape? I felt that any changes imposed by the modern world would only be destructive to this very rugged and beautiful terrain, with its many twisters and sandstorms, unimproved roads and almost non-existent bridges, and deserted forts and settlements in the middle of nowhere. Somehow the austere nature of the landscape and the precious, simple hospitality of its people seemed fundamentally linked to each other. I hardly minded being robbed a couple of times. Nearly everyone wore a gun anyway; those guys just happened to point theirs at me. They took very little, and mainly wanted a ride for a few miles.

In Afghanistan I met quite a few Russians. More important, I met Mark Schmid, who was also traveling toward India. We got along well and decided to continue the trip together. Mark was a medical student from Hamburg, with a wonderful, lively, positive style of living and presenting himself. He could also handle people very well. When he told me I was too rough, and sometimes took things too seriously, I asked him how he could be like he was. His surprising answer was, "I always try to behave as if I were close to orgasm or had just had one." That explained it very well, and I have often thought of his remark and tried with varying success to act like him.

When I reached Kabul, I was completely out of

money. There was still no letter from Christa. Taking the only choice left to me, I drove into the middle of the Bazaar, opened my sunroof, and started offering everything I had to the highest bidder in the quickly accumulating throng of people. Within two hours, all I had left was my suitcase, a rucksack, and my battered, dusty Volkswagen.

The next day I made the acquaintance of the announcer of the local radio station. He kindly agreed to announce the time and place my new Volkswagen could be purchased. I cleaned the car for two whole days without much success and managed to sell it for $350 more than I had paid for it to the Minister of Foreign Trade of Afghanistan, a friendly guy who spoke perfect German and who turned out to be the only person in Afghanistan able to give me a release stamp in my passport permitting me to leave the country without the Volkswagen.

Now I had enough money to make it back to the States, but I had developed amoebic dysentery and was too sick to travel. I shipped most of my remaining luggage back to California so I would not have to carry it. For days I was hallucinating with fever. When I got somewhat better, Mark and I climbed onto the roof of one of the buses to Pakistan for an incredibly scary ride down the winding Khyber Pass road. From our perch on the roof, we could see at every turn the burned-out hulks of buses that had fallen from the road into the deep gorge below.

Our ride ended in Peshawar, where everybody walked around with a gun over his shoulder. The only quarters we could find were in a terribly primitive hotel. We paid less than one dollar a night, but we had to shit somewhere on the roof into overflowing pots while billions of flies covered every part of our bodies. My dysentery had returned, and I had to go back onto this foul roof every half hour, or the effects of the illness would have blown me apart.

We were so desperate to get out of Peshawar that we

traveled to Amritsar on the roof of a train. At least one could breathe up there; the inside of the train was jammed and stifling. The heat and humidity were so extreme that we took showers, fully clothed, every time the train stopped at a station. But Amritsar repaid us for the ride. It was like something out of a fable—beautiful, exotic, and more comfortable than any place we'd been lately. The Sikhs there were kind, and we enjoyed being able to see women again. In Afghanistan we had only seen two—both completely covered from head to foot.

We felt we had to see Kashmir. When we arrived in Srinagar, we were accosted by thousands of aggressive beggars. But it was a beautiful city, and there were lily-covered lakes with houseboats to live on. We rented one right away because I was now so sick I could no longer walk. For two weeks I lay in our houseboat with a high fever and truly terrible dysentery. I exhausted Mark's medical supplies. He kept telling me, "I can't just give you the pills I think might help you. You have to have the will to live."

Eventually I recovered. While we were living on the houseboat, the local merchants came in boats to sell things to us. One was a *shikar,* a hunting guide, and we agreed to go with him on a bear hunting expedition. Two days on a bus and a lot of walking brought us to a tiny village 7,000 feet high in the Himalayas. It was raining like hell. After two days there, the *shikar* woke us up around 3 a.m., telling us, "There are bears in the pear trees." This village had a problem with bears: the Indians had taken away all the guns owned by the Moslem villagers, so the bears were able to eat all their fruits without fear.

I did not enjoy getting up in the middle of the night to solve their problem. It was pitch dark and cold, and I was a coward. We had only one gun, which appeared to be a survivor of the Middle Ages. I was awarded the gun and the first shot, while the *shikar* and Mark had axes as their only weapons. Shivering from cold and fear, we walked under each pear tree, shining a flashlight up into its branches to spot a bear. This struck me as very dangerous, and I expected a bear to attack us. Fortunately for us, we could not find any.

The next day, farther up in the forest, I shot a big brown bear. I fired from an unbelievable distance and hit him right in his rump, and he ran away. We searched for him for three days, climbing over rocks and through thickets, always with our one gun and two axes at the ready. By the end of the second day, the straw shoes of our guide and his friend from the village were so worn out that they went barefoot. Finally we became too exhausted to continue. I promised each of the three tribesmen who had helped us search that I would send a wristwatch to him if they could find the bear, skin it, tan it, and send it to me in California. When I got to Hong Kong, I bought three good-looking, functional wrist watches for a total of five dollars, just in case. Six months later I received a beautiful bear skin. I sent the watches as promised; three months later I received a letter from my volunteers. They had walked for days to Srinagar, hired a scribe there, and dictated a letter to me. Their letter touched me more than any other I have ever received. It started out, "Your gracious grace, Sahib Johannsmeier! We received the watches you promised and will never in our lives forget ..." I cried like a dog, and their letter made me regret less having shot this innocent bear.

After making a pilgrimage to that most beautiful of memorials to a beloved woman, the Taj Mahal in Agra, we arrived in New Delhi. A letter from Christa was waiting, the first in more than three months. There was no check inside—only a letter saying that she had given up our apartment, sold all our furniture, moved to Reno, Nevada, and filed for divorce.

19

Mark had left for Germany from New Delhi. Now Christa's letter dropped me into a pit where there was nothing but pain. I loved this woman. Again and again I asked myself, "Why did you leave her? Why did you make this trip? Why didn't you return when her father died?" One answer kept repeating itself: "I had within me the urge to travel and the belief that by traveling I could make up for the time I had lost. Then, once I had begun my trip, my plan became an imperative, a mission, and I was too stubborn to abandon it." Now I had all the time in the world, but no one to share it with, and no ballast.

I could easily carry everything I owned—a toothbrush, a shirt, a pair of pants, a pair of Indian sandals, and a small bag on a string around my neck. Hidden under my shirt, the bag contained my only valuables—my passport and the money left from the sale of the car. I should have felt free of all obligations, but I was bound to my emotions. Without an audience at home, my trip had lost some of its meaning for me. Christa's letter had made pointless all the plans I had made for what I wanted us to do together when I returned to California and all the promises I had been making to myself about what I would be for her. What hurt me even more than actually losing Christa was my feeling—my certainty—that I had failed as a human being with Christa.

My dysentery came back, made worse by the incredible heat. Not knowing what else to do, I continued traveling through India by train to places I picked almost at random. When I ran into two holy men traveling to Benares and Calcutta, I joined them.

I still have a picture of the three of us standing in front of a beautiful temple. They are only skin and bones. All three of us have beards and have amulets around our necks. We look very much alike. My mother told me later that when my grandmother looked at this picture, she pointed to one of the holy men and exclaimed at how much weight I had lost.

I had always thought of India as the most magical country in the world, a place where people made pilgrimages to find jewels of wisdom. I had imagined myself sitting under trees where great philosophers or religious leaders had once sat. In fact, I received no great revelations in India, heard no great words of wisdom, and had too little time to reach nirvana by living and thinking correctly. I did find wisdom of another kind by observing the country people I encountered in trains and villages. I saw how their close bonds with family and friends made their lives rich and fulfilling, even though they lived at the edge of survival.

My favorite place on the train was up in the luggage rack. There I could sleep or observe the overcrowded compartment below me. On these trains people often traveled with everyone and everything that belonged to them—babies, flocks of elderly relatives, bundles of cloth, chickens, dogs, monkeys, and even pigs. It was a very colorful, ever-changing scene—with terrible sanitary conditions. At every station I got off and took my shower to cool down.

When the three of us reached the sacred city of Benares, I was too sick to see much or enjoy anything. From there I went on doggedly, by myself, and finally reached Calcutta. I began feeling a little better, but saw scenes of some of the utmost misery I have ever witnessed. People were dying on the streets, and vultures fed on the dead human flesh. Trucks came every morning to pick up the bodies. Anybody who did not move was tossed onto the trucks to be dumped I don't know where. I have never forgotten these scenes, from which I learned an important lesson: if you want to stay alive and not be thrown onto the truck, you have to keep moving.

I felt miserable, both mentally and physically, and considered abandoning the rest of my trip and returning to California. But somebody had told me that after I reached Thailand, conditions would start to get better again, so I continued on to Bangkok, where I wanted very much to see the magic temples and the floating market. Suddenly outside a little antique store, I became so weak that I collapsed. The owner, realizing I was very ill, brought me to his home, which was not far away. His family treated me like one of their own, taking care of me for several days while I lay bedridden with a high fever. Their beautiful eighteen-year-old daughter, Gnew, held onto me the way someone would hold onto a favorite dog. She fed me, washed me, petted me, and even slept near me at night. It was so completely natural that the family did not object. This family operated with an incredible sensitivity, serenity, and sense of pride, and Gnew was completely natural, uninhibited, free, and pure. She taught me some Thai words, and I taught her the English ones I most liked to hear.

After I recovered, I took Gnew on a trip through Thailand. I rented a long wooden boat which came equipped with a powerful motor on the back and a tiny older man as guide and steersman. We cruised along rivers and even crossed the ocean to remote temples, lone villages, and isolated beaches. We slept on little elevated platforms right under the sky and ate all kinds of native foods, mainly fruits, rice, fish, and lobsters. It was as close to paradise as I had ever come. I felt like an eighteen-year-old myself, and thought I could stay there forever.

My cash reserves were dwindling, however, and I had no way to earn money in Thailand. I had to get back and face reality. I said goodbye to Gnew and her wonderfully kind family with the honest promise to return. I really wanted to.

Twenty-five years passed before I returned to Thailand. When I went to the area where Gnew's family had lived, everything had been redeveloped. All the buildings were new. I had too little time to locate Gnew and her family, but I remember them with great gratitude and love.

From Thailand I flew to Hong Kong, Japan, and Hawaii, spending only a couple of days in each place. I thought then and have thought more since about what I gained from this trip. First, it gave me time to think about myself: as I traveled, I realized that I wanted to remake myself into a better person and to shake off all the remaining attitudes and suspicions left from my time in prison.

My exposure to the world affected me as much as my introspection. During my trip I enjoyed many periods where I was totally engrossed in the events unfolding, as one can be during a theatrical performance. I carry some of these moments within, like pictures taken with my camera or the favorite poems and books that are part of me. They have become important reference points in my essential self, additional mosaic stones that are part of the basis for my thinking and being.

Toward the end of this trip I had seen many places and scenes, and just being somewhere was no longer enough. I was only satisfied when I could feel I was almost part of the place where I was, not pushed to go on. For example, I hoped I could always remember how simple a life full of love I had been able to live in Thailand.

Before this trip I had been a German living in America, seeing it through German eyes. The trip broke the cast of my national conditioning, making me able to see that the German way was neither the only way nor the only right way. I began to be able to see European, American, and Asian countries differently when I could look at each with knowledge of the other cultures. I found, however, that this new viewpoint on the world set me somewhat apart from most of the people with whom I worked then and later. There was, I realized, some similarity to the way living in a prison society had left me unable to discuss life with people who had only seen it under much less extreme conditions.

Once back in California, I had to return to the remains of the life I had left. To retrieve whatever was left of my clothes and possessions I had to find Christa. She had given no address or telephone number in her letter or on its envelope. I decided to fly right to Reno; there I called every hospital in town several times until I found a person who gave me her address. I went to Christa's place in the middle of the night, and found that we still had a bond between us; after just one day, we had connected again. I loved her more than ever. And I was certain that, as a result of my recent experiences and self-examination, I had become an easier person to live with—more easy-going, more freed from my past. I now tried to live my life with an attitude like that which Mark had demonstrated.

I had to begin earning money without delay. I bought an old VW, drove to Palo Alto, rented an apartment, and showed up at Hewlett-Packard the next day. I found that I still had a job, and that two patents had been granted to me during my time away.

When I went through all the mail that Christa had collected during my absence, I found several margin call letters from brokerage houses and banks. When I visited these institutions, I found that all my stocks except the ones that were totally worthless had been sold. I was more than broke: I owed money to the bank. When I mentioned my situation to some older and financially more experienced friends, they told me all this was a good thing—it was a valuable experience at such a young age, and I would never forget the scars it would leave on my rear end.

Indeed, I thought long and hard about this experience and made sure that I understood how numbers and percentages could work for or against me. The experience may have taught me useful lessons, but I found it hard to face the reality that I was starting from zero again.

Christa visited me several times on weekends. We still loved and needed each other, and it seemed to me that now that we were separated and living in different apartments we got along better than we had when we were married. Christa told me later that on the day set by the Nevada court to officially terminate our marriage, she had waited for me on the steps outside the courthouse, hoping I would appear and we could reconsider.

But I had not shown up. I had ducked into the foxhole of my mind and let the divorce tank roll by. I had decided to live my life free of entanglements. Still, Christa and I stayed friends for another year, until she found somebody willing to commit to her—the guy who had wished I would not return from my journey around the world.

Christa probably needed someone less dominating, intimidating, and bohemian than I, as my mother had needed someone less bohemian than my father; but I have never felt that I handled my relationship with Christa well. She was very attractive, had a fine character, worked hard and efficiently, took care of everything neatly and conscientiously, and was a super companion and comrade. Our relationship never functioned smoothly, though, perhaps because we had different backgrounds, expectations, attitudes, and ambitions. Also, we were both experiencing the stress of trying to make lives for ourselves in what was still for us a foreign country. I felt that her outlook and aspirations were too cautious and limited, tending to keep me immobile when I needed to take off and fly. We also had some horrible fights about her smoking. But all these problems could have been overcome had we been a better match for each other.

After my divorce I never admitted to anybody that I had been married. I handled the matter as my mother had handled my father's death. I could never talk about the relationship, or my behavior with Christa. I did not want to break the sheet of ice I had allowed to form over the emotional landscape of our relationship. I had willed into being the same sort of ice sheet over my prison experiences, and over my feelings when I left East Germany; I did not want to reexperience those traumatic sensations.

During that year when my relationship with Christa was coming to an end, I made some very good moves at Hewlett-Packard. Just after my return, I followed my boss, Ed Phillips, to H-P's Advanced Research and Development Laboratories—the place where most of the new projects began which would eventually strengthen existing divisions or become the seeds from which entire new divisions would sprout. With literally hundreds of ideas and projects to work on, it was a paradise for creative people. At first I worked helping Ed, but when he decided to concentrate all his efforts on one ultrasound project, I went to the department manager, Paul Stoft, and told him I wanted to take on Ed's job of managing all the mechanical engineering in the department. I said, "I believe I can do this job well, and perhaps do some things even better than Ed. For example, I will utilize talent better." Surprisingly, Paul bought my pitch and started assigning engineers to me until all the mechanical engineers in the department worked for me.

I had reached very high in taking my new job, and I

had to work very hard. As Hewlett-Packard hired only the very best engineers, I was managing people of very high caliber. We had plenty of very promising projects and ideas to work on, and the company structure created by Bill and Dave gave us considerable freedom and encouraged us to use our talents to the fullest.

One day a new employee arrived in our department, carrying a cigar box with some function keys on its top. While working at a mechanical calculator company, Tom Osborne had become convinced that mechanical calculators now could and should be replaced by electronic ones. The company he worked for had rejected the idea because it promised to make its own product line obsolete, but Bill and Dave thought his ideas were worth pursuing. I was the first man assigned to the "cigar box" project; immediately I hired an engineer who had experience in the design of keyboards. We projected that we could sell as many as 200 electronic desktop calculators a month. In fact, Tom Osborne's cigar box was the beginning of Hewlett-Packard's multi billion-dollar desktop and hand-held calculator division. It would soon sell tens of thousands of units, making our beautiful slide rules obsolete. Not knowing the future, we worried over our decisions. I still remember Barney Oliver, H-P's Vice President of Research and Development, coming around and testing the feel of the newly developed keys, as we pondered which functions to provide and whether we should use RPN (reversed Polish notation) in the calculators' operation. We also worried about whether the huge multi-layer printed circuit boards would work: they were an exciting innovation then in 1965.

During this period, about two months after Christa and I had stopped seeing each other, I noticed an unusual woman arriving at a party. She looked like a fresh peach. Her elegant clothes and manner epitomized high refinement and class. She was accompanied by a handsome man and seemed almost too beautiful to talk to. Even so, I walked right up to her and introduced myself. We never stopped talking that night. This South American beauty, whom I will call Estella, came from a family proud of its pure Spanish descent. Her father had been ambassador to Spain. Estella had just arrived in San Francisco and was trying to break away from her family, which, in the manner of most upper-class families in Latin America, had kept her strictly confined. Even though she was twenty-four years old, she had never in her life been allowed to do such simple things as walk barefoot or kiss someone. To me she seemed a precious natural resource, perfect in every way and painfully sexy. I knew I was reaching a little higher than I thought I could achieve, but I put everything I had into courting her. It worked; I soon eliminated the handsome guy she had been with.

Estella came to love me with all her considerable heart. She called me "Baby, baby, my baby"—but then she also said that to every animal she encountered. She converted me into a gentleman, teaching me those good manners I had not learned from the movies I had seen as a kid. She did not have to push me into trying to make it big—I was already working on that feverishly—but she gave me new encouragement to succeed. I became a living example of the saying that most men can do little for themselves but everything for a woman.

While she made me more alive, however, like Medusa in reverse she would from time to time turn herself into stone. About every three weeks, a severe psychosomatic problem attacked her with pain so great that she became totally paralyzed, remaining in bed for a week before the pain went away and she could return to being the happiest and funniest person on earth.

During our first two years I was able to handle this condition very well. I made her jump up and down, or walk barefoot in the grass, or splash through the water at the edge of the ocean. Such techniques jolted her out of the stone-like stage. But when she came under the influence of a psychiatrist, he turned her against me and she reverted to severe illness.

I really wanted to marry Estella, but I was always afraid that her problem would get worse. On a trip to Italy I became acquainted with a psychiatrist. He and I walked the beach each morning for several days, discussing Estella's case medically and philosophically. We could not avoid a grim conclusion: Estella would only get worse. If, like many sick people, she ended up blaming her problems on those who lived with her, I would eventually get all the blame.

At this time I was keeping myself very busy. I was taking some courses in electronics and studying toward a MBA degree at Santa Clara University. I found that I knew spoken English well enough, but my lack of experience and skill in writing English made it almost impossible to do well in some of the courses. Then, in addition to my work at Hewlett-Packard, I was also giving time to developing several inventions on my own. One was simple—a photo album page that would hold snapshots without tape or photo corners. But there were already many similar products, well established in the market, and I could not obtain financing.

During a three-thousand-mile trip through Mexico, I took more than two thousand photographs. Everybody who has taken a lot of slides has to spend boring time in storing, sorting, and selecting them in order to show them. Dealing with my Mexican slides gave me a great idea for a slide projector. I invented a projector that uses an album page as its magazine. A built-in light table enables the operator to preview up to thirty-six slides while selecting any of them to project onto a screen.

I spent a whole year of my spare time designing and building this projector. It was good enough to make my boss Paul Stoft say, "You will be the first engineer in the entire lab to become a millionaire." Manufacturing the projector, though, required an investment of hundreds of thousands of dollars to pay for tooling and building the necessary production line. In California in 1965, it was far more difficult than it is now to obtain financing. Silicon Valley was not yet hatching promising new companies at a great rate. No one was talking about financing companies with the aid of venture capital firms. With many slide projectors well established in the market, I could not obtain financing. My only choice, then, was to offer the projector to companies already in the business.

Several companies invited me to come demonstrate the projector, but only one man, the CEO of a photo-optical company, visited me in California; he flew out in his private jet. He loved my projector and invited me to fly back to Chicago with him, see his factory, and visit the chief buyer at Sears, Roebuck; then we would continue on to New York to meet the principals of some other marketing organizations. I jumped at his offer, took some time off, and off we went in his jet. It had a pilot, but the CEO liked to fly the plane himself. He even allowed me to sit in the co-pilot's seat and pretend to fly it a while; for safety, however, the plane's automatic pilot was on.

The chief buyer at Sears liked my projector very much, but he said he had doubts whether the untrained sales clerks in his stores could handle selling it. When we visited an important marketing organization in New York, a Jewish man who also loved it told me in German that "the life of an inventor is a very difficult one."

I was fascinated by the lifestyle of my new friend. He pushed himself hard but lived life to the fullest, with generosity. He had a fantastic penthouse right on Central Park and seemed to know every fashion model in town. He appeared to be in command of all aspects of life, and to know how to handle even hopeless situations. When a traffic stoppage had blocked our limousine on the way to the Chicago airport, the CEO had ordered his chauffeur to drive along the shoulder, with the right wheels on the grass near a ditch. We drove more than ten miles this way until the freeway finally became clear, and we did not miss our appointment in New York.

During our return flight to Chicago the CEO asked me, "What do you think about the reactions to your projector?"

I said, "I think they are very positive."

He surprised me by saying, "I don't think so. It is a maybe."

Two weeks later I visited another optical company in Boston. They got very excited about the projector and offered me $5,000 for a three-month option; during that time they would keep it and look at it more closely. I accepted the offer. That amount paid for all my expenses on this project. More important, receiving the first money I had made working for myself made me feel very happy and hopeful.

I had been pushing myself hard, keeping all these balls in the air. I now felt exhausted. I knew I had to get some rest. I wanted to go to a beach and sleep for a few days. Recently I had heard a couple talking about a place called "Club Med." They said the one they'd been at, in Guadeloupe in the West Indies, was great. With my $5,000 burning a hole in my pocket, I took a plane there. At eleven o'clock at night I arrived at the Club. Nobody was at the reception desk. I left my suitcase there and followed the sound of loud music. A half-crazed crowd was jumping up and down, singing "Let the Sunshine In," from the musical *Hair*. Everybody knew it. I joined in. I stayed there five days and never slept a single night. Sleep was not what I had needed. Club Med re-ignited the immense power and dynamic of free life in me.

When I came back, refueled and happy, a friend who

had a business in climate control equipment asked me whether I could build him a particle counter. I said, "I can build anything." Then I asked, "What does it have to do?"

"Well," he said, "it has to count and categorize particle pollutants in the air. The semiconductor industry needs absolutely clean air in the rooms where computer chips are made. The machine should signal when any pollutants start leaking in."

I promised to develop it in my spare time in exchange for a royalty and the payment of all my expenses. Now I had something to work on while my projector was being taken apart and analyzed in Boston.

Meanwhile, I had many projects at H-P to oversee. We were developing ultrasound devices that could look inside the human body; lasers emitting coherent light which could be focused more precisely than light from other sources and used to make holograms; blood cell counters which could quickly count cells and sort them into many categories; an oximeter that could measure the oxygen content of one's blood without piercing the skin; precursors of today's printers and calculators; mark sense readers that could sense pencil marks and convert them into electronic numerals; and a hot scalpel which could cauterize small blood vessels as surgeons operated. Most of these projects eventually produced successful products.

From my work at H-P I learned a great deal about what products should and should not be developed, and about running projects, managing people, and using other resources. An important principle of administration that I always tried to practice was that a problem had to be defined carefully, and the responsibility for it assigned to someone clearly, so that he knew he had it and that he did not share it with somebody else. I had learned this principle years ago in school: our teacher told a group of us to solve a mathematical problem. No one came up with a solution because each of us thought someone else would solve it.

All through this period Estella's sickness kept getting worse. Finally she decided to return to her family in South America. I thought this was a bad idea, since even she admitted her parents were the source of all her problems; but I did not know how to help her. I had made up my mind not to marry her, and reluctantly I let her go.

The Boston firm returned my projector; they had decided not to acquire it. Their decision was a big blow to my hopes. As I thought about it, I realized it was also a warning of just how difficult it would be for me to succeed with a product for the consumer market. Hundreds of thousands of people had specialized in the inexpensive manufacture and distribution of consumer goods, while my experience had been in designing and developing technically innovative instruments. If I tried to compete with consumer market experts, I would always be having to learn what they all knew already. From now on, I would stay on my own turf.

Fortunately, I had the particle counter project. I had already filed for a patent on the design and had the principal parts working well. Three months later I had it all together and it went into production. Unfortunately, the company's management changed, and I had to deal with a man who did not know or want to know about the verbal agreement I had made with the owner. He wanted to get rid of me and milk the product for the company's benefit and his own. I could not work under such conditions, and we settled the matter for $36,000. This was three times the salary I was making at H-P, and I had earned it in nine months in my spare time. That made me think.

I wanted a break during which I would have a chance to re-think my future. I also had a desire to visit Estella—

which would allow me to see South America as well. Hewlett-Packard again kindly gave me a six-month leave of absence but said they could not guarantee I would get my position back when I returned.

Shortly thereafter, I headed south through the countries of Central and South America. I did not travel as frugally as I had in going around the world, but I always stayed in reasonably priced hotels. I found I met more interesting people there, partly because it was harder to make acquaintances in the expensive hotels, and partly because reporters, writers, and artists usually couldn't afford high-priced accommodations.

When I reached her country's capital, I called Estella. She was delighted to see me, and I realized again how dear she was to me. We went to many social events, of which the most memorable and colorful were at the foreign embassies. Her father knew everybody, and I was introduced to ambassadors from every nation imaginable. By then I had learned quite a bit of Spanish and French and could understand and speak some of the polite phrases exchanged on such occasions. With Estella at my side, I felt right at home in this environment.

We went often to the racetrack. Racing here was much more elegant than in the States. Everybody dressed up, and everything had an air of importance. I bet a few times, never winning or losing much. While I enjoy taking risks where my skill can make it possible for me to succeed, I never much liked betting on horses or gambling in casinos. My instincts told me that without inside knowledge, one would ultimately lose one's money.

Estella lived with her parents in a big mansion in a section of the city where the wealthiest and most important families lived. I was fascinated by observing the organization of that household. Estella's mother managed the butler, and he then managed the staff. Everything operated by well-defined rules, like the mechanism of a clock work. At dinner Estella's mother would ring the powerful little bell placed close to her to set in motion the ceremony of servants bringing and serving food and wine. I felt as if I had entered the world that existed in the movies of my childhood. How formal and polite they were to me, and even to their own son and daughter; how much energy each seemed to expend just maintaining his or her own image; and how strange everything appeared to me, knowing from Estella as I did that her parents fought often in deciding which small details of expense were necessary to maintain their prestige in their world and prevent any decline from their high place in society. It made me understand why they had kept Estella in a social straitjacket, and how these constraints may have been the source of her illness.

When it was time to leave, I thought about asking Estella to come with me, but I did not have the strength. I don't know whether she would have come.

Some years passed before I saw Estella again. I learned that shortly after I left, she had returned to the hospital and had been given electric shock therapy. She seemed to be well, but I felt as if her essential qualities, her warmth, blitheness, and sense of humor which had made her so adorable, had been destroyed: she could not laugh anymore. I have often wondered whether, if I had asked her to leave with me and she had assented, I might have been able to save her from that fate.

Instead, I flew on to Peru alone. After visiting many of the ancient sites of the Inca civilization, I traveled to the jungle city of Iquitos on the upper Amazon River, a hiding place for asocial elements and individuals unable to show their faces anywhere else. While eating and drinking in restaurants there, I met many colorful characters, all friendly, approachable, and essentially harmless. Most had nothing to do and all the time in the world for conversation and hatching plans for new questionable activities. Many of them were lonely and carried pets, mostly tiny pocket monkeys, smaller than a hand, which were tied to their wrists with little chains or pieces of string. In the evenings, the whole place swarmed with prostitutes who cost almost nothing. I was approached each day by some who wanted to do it for absolutely nothing.

One day I happened to meet three nuns who were missionaries to Indian tribes whose villages were seven days farther up the Amazon River. I talked them into taking me with them when they left Iquitos. The trip turned out to take much longer than seven days. I saw endless amounts of river water and jungle. The Indians were more than a little shy; when I tried to photograph them, they got angry or frightened and scratched my arm with their fingernails hard enough to draw blood.

After leaving the jungle, I traveled from Ecuador to Chile and through Argentina, Uruguay, and Paraguay to Brazil. On the Copacabana in Rio de Janeiro I met perhaps the only two Jewish ladies who were grateful to Hitler. They had settled in Rio after the war and loved it. They said laughingly, "We thank the great leader (the *Führer*) for throwing us out of Germany and 'leading' us to this wonderful place." I found that to be a curious way of looking at life.

While I was in Rio de Janeiro, I started feeling generally unwell, and developed a severe pain in my back and side. I tried to ignore my illness; I moved on to Brasilia and then to San Salvador Bahia, where I looked into the faces of a dozen murderers whose heads had been preserved in big glass jars to be exhibited forever at the local museum. Each day I felt more tired, and each day the pain in my back and side increased. I saw a doctor; he told me I had the flu.

I took a flight to Barbados. At one of the wonderfully white beaches, I almost collapsed. I had to see a doctor immediately. As I stepped through the door of his office, the doctor looked at me and said, "You are a dead man!" He was nearly correct; I had a very advanced case of hepatitis. My skin and eyes were yellow, and my urine was black.

This man saved my life. He immediately put me in a hospital, where I remained for four weeks. Then I was able to return to Palo Alto, but I had to spend another four weeks at Stanford Hospital. I was never able to determine how I had acquired this infection, whether from the strange foods I had eaten, such as barbecued bowels, or from the scratches I had received from the Indians. I knew I was lucky to have survived; eventually I recovered fully.

While I was ill, I had plenty of time to think. I thought about my South American trip, and other trips in the past. The ones I had enjoyed most, I decided, had only lasted about three weeks, and had been with a beautiful woman. I remembered especially two recent trips to Tahiti with Terry, a girl I had met at the pool of the apartment building where we both lived. She was fifteen years younger than I and had the most beautiful big brown eyes. To attract her attention, I had jumped headfirst into the pool from the roof of the three-story building. Fortunately, I had my arms in front of my face when I hit the bottom of the pool, but my head bounced forward and through my hands, hitting the concrete and knocking me out for a moment. When I came to the surface, I was bleeding like a pig, turning the pool water into a red sauce around me. Holding the cut tight with my hand, I ran up to my apartment to look in the mirror and clean the wound. Terry knew I had done this stupid stunt to impress her and followed me.

We became good friends and later lovers, even though we were in many ways mismatched. Terry was very loving and very easy-going, and we often behaved like children together, laughing at everything. Although she was very smart, Terry made me laugh all the time by playing the innocent and saying the most amazingly naive things. Once, when I invited her to a concert, she replied, "To a concert?" She wrinkled her nose: "With music?" When I told her she was lazy, her answer was, "How can you say something like that? You have never seen me work!" When we admired two huge bulls lying in the grass out in the country, I threw a little rock at one and they both got up and walked away, showing their big balls. Terry floored me by asking, "Which one of them is the female?"

I called Terry from my hospital bed in Barbados and told her I had a bad case of hepatitis and had to take care of myself since I had only this one liver. Even though she was very concerned about my condition, she could not resist commenting, "My God, I did not know you had only one liver." This almost killed me: when something is wrong with your liver, laughing is very painful.

While I was in the hospital, I also developed and assessed a number of scenarios for my future. I could just continue to hold a job. I could live the life of a bohemian lover, frequenting parties and drinking lots of beer and wine. I could give all my time to being an inventor. I could become a serious painter, or a writer, or make movies. I came to the conclusion that I could not yet paint well enough to make a living; that I didn't know enough even to begin to make a movie; and that any book I wrote about prison, bombing, drowning, cracking ice, avalanche, dysentery, and hepatitis would be boring. Perhaps the hepatitis contributed to my gloomy view of life, but I felt dissatisfied with my status as I neared the age of 40.

20

Early in 1968 I went back to work at Hewlett-Packard Laboratories. To my surprise, I had not been replaced. Soon I was back in the same old routine of directing and guiding engineers in solving problems and designing what was needed for the various projects. But I was less interested by the work. I noticed that when my engineers came to ask me questions, I would start to yawn—even as early as ten o'clock in the morning. Something was very wrong: I had already had plenty of coffee, and I was well rested from my long vacation and my hospital stay.

I sensed that I had reached a dead end at Hewlett-Packard. As a mechanical engineer in a company that made electronics products, I could see no position for me higher than the one I already occupied. When I thought about money, I figured out that even if I stayed a bachelor and lived frugally, I could save only $5,000 per year. Saving at this rate for the next twenty years, I would still not accumulate enough money for a comfortable retirement.

When H-P held an open house for employees' family members and friends, I took Terry to the Lab to show her where I worked. My space was like everyone else's—a desk and a work bench. Astonished and disappointed, she said, "This is it? This is what you're dressing up for so neatly every day?" This was one of the most important comments anyone ever made to me. It turned my simmering discontent to alarm. I knew then I had to do something different: something that had potential; something big; something on my own. I had to stop working for wages and become an entrepreneur.

I could have found a thousand reasons not to go

ahead with this change. Among the most frightening ones were that I had no experience starting a business; I had very little money; making the business viable would require a long and continuous commitment of my time; and running it would force me to carry responsibilities I had always avoided. Obviously, I would have to sacrifice my travels and spare-time activities.

I had no idea that I would also become my own janitor, bench builder, purchasing agent, engineering department, manufacturing director, advertising manager, patent attorney, director of sales and marketing, chief executive officer, and chairman of the board. I had not realized that entrepreneurs have to take on all these roles themselves when there is no money to hire staff, or when people hired to do the jobs leave.

I tried to visualize the tough and very serious undertaking ahead. I was worried about my lack of experience in business matters. I had only a highly simplified, pieced-together mental model of how a business functioned: one had to create a product, manufacture it, sell it, pay all the bills, and have a profit with which to enhance the product or create a new one, to continue the process.

Still, I thought I had a good chance of making it. While I had never started a business before, all the jobs and experiences I had had in my life had been training of a sort for this. Of course, this training had all been with wooden bazookas: I had not yet faced battle with real weapons and real enemies. But I realized that almost everyone starts any new undertaking without experience and that no amount of training with theories and models can ever fully prepare one for the real thing.

I was as ready as I would ever be, and there was no excuse to wait any longer. Remembering again and again Goethe's famous line from *Faust*, *"Am Anfang war die Tat"* ("In the beginning was the deed") I made up my mind and quit my job.

I rented a space on Old Middlefield Road in Mountain View, designed some work benches out of wood, ordered the boards cut to size, and nailed them together. A few days later, I bought some used desks, used drafting boards, an old lathe, a drill press, and a milling machine. Then I glued a sign on the door saying "Opticon." This, newly printed business cards, and some scattered drawings and parts from my slide projector project made the office look like a going concern. With nobody working for me, Hewlett-Packard's projects behind me, and not even a product to worry about, I had entered a wonderfully virgin landscape and felt free. Even though I had not defined precisely what projects I was going to work on, I felt confident that once I did, I would be able to find smart people to solve the problems. The Bay Area was full of them, and I had learned how to utilize their talents. If I had an interesting project and a work place that looked like what they were used to, most would make themselves available, even if I had no cash at hand. They would work part-time, as I had done on the particle counter. Americans like challenges and making money, and would rather work than take time off for vacation.

Now I needed to find a project on which I could bring to bear all my knowledge, experience, inventiveness, tenacity, and understanding of people. The fact that I did not have any such project in mind did not bother me. I decided that I would approach entrepreneurship as I approached traveling: on my trips, I had always traveled without reservations or detailed plans. I knew what I needed to do first: pick the right undertaking and figure out roughly how long it was going to take. Then I had to be always ready to change my plans as circumstances changed.

Even if I didn't find a project at once, I could take in design work and help other companies develop products. Or I could start to design and manufacture an even better particle counter than the one I had already designed. I could stretch my cash resources by hiring people on a consulting basis and cutting them in on the deal at hand.

I felt enthusiastic and full of energy. I was not yawning any more, even without coffee. I had kicked myself into a mode where I had to perform. I knew that I worked at the peak of my ability under pressure and that I had great tenacity at carrying a project through to completion: also, nothing stimulates these traits as much as working for oneself.

A few days after my office was set up, I went to a big electronics show to examine existing power supplies and other components I could incorporate into a product. Anything I could buy ready-made rather than develop and manufacture myself would save me valuable time and cash.

I noticed that many of the displays at the show were connected in one way or another with the semiconductor industry—a new, very hot business that had not even been mentioned when I was in engineering school. The name over one display caught my eye: Kasper Instruments, Inc. I knew that name because I owned some of that small company's stock. Its owners, vice presidents at Hewlett-Packard, had sold it to me when the stock's price was low so that I would have an interest in the company if they wanted my advice and ideas.

Kasper was displaying its product, a very sophisticated piece of equipment that the emerging semiconductor industry needed in the manufacture of electronic devices such as transistors and integrated circuits. These devices had already found their way into transistor radios, electronic power supplies, and other products with rapidly expanding markets.

I knew that these electronic devices were processed

into very flat and polished semiconductive silicon crystal plates called "wafers." I did not know much about the process by which these devices were made—the manufacturers kept that secret and proprietary; but I did know that the process was extremely difficult and complicated. A number of devices were made on one wafer at the same time ("batch processing") by processes which included such steps as spinning a photoemulsion coating onto the wafer, baking it, exposing and developing it (as in photography), etching extremely small channels into the wafer's surface with acids, doping (implanting ions into wafer surfaces to increase or decrease their conductivity), and using a metal vapor deposition process to create all the many connections. I also knew that a method called "photo masking" played a very critical part in making it possible to process small selected areas of the wafer's surface, and that Kasper's equipment was needed to do it.

Kasper's equipment appeared to be well designed. In a 36-inch-by-36-inch space only 14 inches high, it held several high-powered microscopes and illumination systems, several mechanical x-y stages, a high-powered exposure system, micro-manipulators, many castings and machined parts, pneumatic valves and pneumatic logic blocks, electrical solenoids, circuit boards, wire harnesses, and many electronic components.

While the machine was very functional, many aspects of it appeared to me to be very rudimentary and in need of improvement. Its most obvious shortcoming was that it had to be operated manually. On semiconductor manufacturing lines, the Kasper equipment operators had to peer all day through high-powered microscopes to laboriously and very meticulously bring extremely tiny windows (1/10th as wide as a human hair) on two different objects into perfect overlay alignment (1/50th as wide as a human hair) before instructing the machine to print onto the wafer. Getting the alignment right each time all day long required incredible skill and patience, and was very demanding and tiring. In fact, no one could get it right every time; and when the alignment was wrong, the devices produced had to be thrown away.

As I looked at this piece of equipment, I immediately saw ways to improve it. I thought it might even be possible to automate its operation fully, thereby eliminating the necessity to have operators. I also knew from my particle counter project that human operators were a source of air-borne particles which could ruin the devices being produced. These problems were nasty, but intriguing: getting rid of these causes of bad device yield could improve output quantity significantly and justify a much higher price for Kasper's equipment.

I kept thinking, and realized how exciting a project this was and that the skills required to solve these problems exactly matched those I had acquired from my education, experience, and interests. There was a very difficult job to do, a major contribution to be made, and perhaps a large reward to be earned.

The next day I called John Young, Kasper's chairman of the board, and told him what I thought I could do. He brought me together with Bob Johnson, the president of Kasper Instruments, and we worked out a deal whereby my new company would take on the design of various improvements for the existing Kasper unit.

Now I was in a position to hire people and to develop, design, and buy parts and services, charging everything to Kasper, including all salaries, rent, and office expenses. It was a compromise and not what I had originally planned. But my new venture was up and flying in a matter of days. It was exhilarating! Suddenly I was doing everything I had been doing for H-P, but for myself. Why had I not done this much sooner? I just had not been ready and had still been obsessed with making up for my lost time, the years spent in prison. But now I was all caught up and moving fast.

I hired several very good engineers and technicians, some of whom I had worked with at H-P. After I showed Kasper my first results, the company began turning over to me all its development and design projects, and also transferring to my company the employees who had been working on them.

Within a year I developed several inventions in this new field. The first one was a new method of printing the patterns onto the wafers called "proximity printing." It saved the hard-to-get and expensive photomasks used in the process and increased the percentage of devices that worked. Around this new principle we designed a whole new piece of equipment which also included a newly created automatic wafer-handling system. The new equipment worked excellently and sold for more than twice the price of Kasper's original model.

Then I invented the first step-and-repeat equipment capable of printing and overlay aligning onto wafers. It used reduction optics to align and expose the fine windows onto wafers directly from the glass plates, called "reticles," which had previously been used to make the photomasks. It completely eliminated the photomasks, which were very expensive. By projection-printing with high resolution, we could make very fine lines (1 micron width) and overlay them precisely. There was not much need at that time for high throughput, and these machines were mostly used in research labs or to make microwave devices. But their projection printing with 10:1 projection optics could deliver the finer resolution and the increased overlay precision (alignment) that were prerequisites for making the next generations of devices.

When Ed Phillips came by one day looking for work, I immediately offered him a job and was happy when he agreed to join us as a consultant and to work with me on this project. I even was able to get my former H-P boss,

Paul Stoft, to lend his expertise as a consultant. With this powerful team, we invented a new device that enabled the machine to align automatically, thereby eliminating altogether the operator and human error.

This made the machine so much more efficient that it catapulted the price for Kasper's equipment to more than five times the original amount. In this new field of manufacturing microchips, countless other innovations and improvements like these were needed: increasing yield, saving photomasks, processing bigger and bigger wafers and bigger and bigger devices such as integrated circuits. We did a lot of inventing, filed numerous patent applications, and obtained many patents.

All these new projects pushed my particle counter project into the background. I lost the chance to cash in on the huge smoke detector market, which used an inexpensive particle counter that someone developed later. Instead, I took all my savings and bought Kasper stock. I had no idea at that time how dangerous it was to bet everything I had on one company.

By late 1969, I had become Kasper's Vice President of Research and Development. It was my responsibility to come up with product improvements, as well as new devices that would satisfy the ever-greater demands of makers of microchips, thus solidifying Kasper's place as a source of the equipment these companies needed. I held the fate of Kasper in my hands: my work would make the company succeed or fall by the wayside. Accustomed to taking risks, I was comfortable with this situation, but I should not have been. My company was totally devoted to and dependent on Kasper.

In 1970, when the economy went into recession, the semiconductor industry was hit badly. Bob Johnson and the company's owners, the vice presidents from Hewlett-Packard, talked me into merging Opticon into Kasper in exchange for options on more shares of Kasper Instruments. After I agreed to this merger, I was made second in command to Bob Johnson, Kasper's chief executive officer.

Until then I had no idea what a heavy burden of responsibility an executive of a firm carries. With someone else responsible for making decisions and for their consequences, I had always been working under a protective umbrella. But Johnson folded up that umbrella immediately by telling me, "This company's fate is going to be determined by you and me. We will make it or break it. Everyone else is just part of the mechanism, and we must employ them as effectively as possible." His words made me suddenly aware of how serious this game was going to be and how much he and the company would rely on me.

I liked Bob Johnson. He was smart and exuded confidence. He was well-educated, with an MBA from Stanford, and knew a lot more about how to run a business than I did. Even though he was not that much older than I was, I saw him immediately as my mentor; by emulating him I could perhaps learn right here on the job what I had missed by not completing my MBA.

When we began to work together, I observed with great satisfaction how tightly he ran the organization, and how he always found economies and saved money for Kasper Instruments as my grandmother had done in her small household. Really, his efficiency was unbelievable; for instance, he had found Gloria Fields who did all his secretarial work and also served as an administrative assistant to him. Until the company grew quite large, Gloria also did all the accounting. She could do anything and was like a watchdog shielding him. She would have died before she told anything even remotely confidential to anyone. Only she and he knew all the important numbers—how much cash was on hand and how many units had ever been shipped.

Everybody followed Johnson's instructions precisely. He could take notes better than a court reporter, but wrote them in such tiny handwriting that only he could read it. He hesitated and thought before answering his phone to make sure he spoke the right words with the right tone to the person at the other end of the line. He could talk sweetly and authoritatively to customers or wring the best deals out of suppliers and the best service from our representatives. He understood how to carefully prepare and then present our performance figures to our board of directors. He always covered all the ground before he would do anything.

Because Johnson was so smart, did everything so well, and had silver-white hair, some of Kasper's employees called him "the Silver Fox." But I often called him "God." He was, as many chief executives are, the personification of the company to employees, customers, suppliers, and banks.

Having come from a much larger corporate climate and environment, I found Kasper's much more transparent. Often I felt as if I were in a theater, able to observe every part of the stage and every actor, judging the play and the performances and seeing where all the weaknesses were.

Bob fascinated me. Different as we were, we developed some sort of friendship that helped bridge the huge gap between a very well-educated, seasoned, and patriotic American who liked total control and security, and who had grown up and studied in a profit-oriented world, and a somewhat artistic immigrant inventor, by many measures less well educated, with borderline hippie ideas and a background in a socialist planned economy where the very idea of making a profit was absent or condemned.

Bob and I never fought or had arguments. Everything between us was considered and solved with logic. I wish

I could have obtained a description of me at that time from Bob. He must have observed as much about me as I did about him. I discovered his special traits: how he always studied every item on the menu in a restaurant with great care, and then with great difficulty finally decided what to eat; how his voice took on a different sound when he thought he was being especially clever, had a great new idea, or did not want to disclose something; how carefully he thought everything through, considering every imaginable eventuality before making a decision.

That was Bob's way: he needed to be prepared, and then most of his decisions were exactly right. Perhaps because I was a little younger, with younger attitudes, fewer burn scars on my butt, no family and less to lose, I had more willingness to take risks. I thought that I, the immigrant, had more of the pioneer spirit. I could never have prepared myself as well as Bob did—it would have taken the fun out of everything, giving me the feeling that I had already done what I was about to do. It would have been like reading too many travel brochures before embarking on a trip. But I adopted many of Bob's habits, waiting a few seconds before answering the telephone and thinking longer before I spoke.

By being so much closer to the battle line than I had been at Hewlett-Packard, I learned a great deal about business ethics—that they could never be the same as the ethics I knew from life, family, friends, and my reading of poets; that in business almost every decision was colored or dictated by necessity; and that if things became really tough and the company's existence was on the line, one did what one could get away with. I learned how to hire and fire people, how to delay payments, how to price and how to sell with discounts, how to negotiate contracts, and to whom one could sell equipment that was becoming outdated.

I always admired Johnson when he put on his charm

show to court customers. When we traveled to Japan, the Japanese always raised the American flag to receive him respectfully, wined and dined us royally, brought us to the best geisha houses, and seemed to bow their heads farther for him than they did for their own superiors.

I never knew whether Johnson would have been a friend of mine had it not been for Kasper. I suppose not; he was probably too cautious to give himself unreservedly. But we complemented each other perfectly and worked and fought together well, traveling together to all the trade shows and visiting customers all over the world, selling our equipment and building Kasper's fine reputation. Only once, while we were soaking together in a big bathtub in Kyoto, Japan, after we having been entertained by geishas singing, filling our cups with sake, and even faking making love to us, did we become close enough to each other to talk about private things, such as family, girlfriends, and the thoughts and wishes we harbored for the future—which life rarely fulfills.

After one such geisha party, I nearly killed Johnson. We had paid no attention as the geishas kept filling our glasses with sake, and had no idea how much we had consumed until we tried to walk down the stairs to where the limousine was waiting for us in a snowstorm. Johnson sneaked into the back seat, and I sat beside the driver. It was snowing and blowing like mad, and we had sixty miles to cover to get back to Tokyo. As soon as the limousine began moving, I felt that everything was about to come up from my stomach. I concentrated immensely hard to control my nausea, and opened the window to get the help of fresh air. For the first few minutes Johnson kept protesting bitterly from the back, but I told him that if I closed the window even partway, my stomach would empty itself into the limousine and Kasper's fine reputation would be blemished forever. Johnson stopped complaining after a while—he must have fallen asleep. I held onto myself while the snow kept blowing through the window. When we arrived in Tokyo two hours later, I had not vomited into the car, but in the back Johnson was covered completely with two feet of snow. It was a wonder that he had not suffocated, but he did come down with a terrible cold that could have killed him.

In 1973 at a trade show in Munich, Germany, some gentlemen from Jena, East Germany, stepped into our booth. They were very interested in our equipment, wanted to buy it, and asked me whether I could visit Carl Zeiss and give a talk about the equipment. I told them, tongue in cheek, that they would have to send me an invitation signed by the president of their country that guaranteed my safe return to the West. They assured me they would do that, and I actually received an invitation stating that they had had my files checked and found that I had really done nothing; however, it lacked the signature of the president of the German Democratic Republic. Of course, even if I had trusted them, I could never have accepted an invitation: our equipment and technology were restricted from being exported behind the Iron Curtain. Also, I could never have trusted them to allow me to return to the West.

During my years at Kasper Instruments, we were operating in a dynamic environment. As the newly invented geographic term Silicon Valley became almost synonymous with the semiconductor industry, new companies started, grew, and burst into flower almost as fast as the plants seen in time-delay nature movies. Constantly changing technology created new needs almost daily. We engineers had to come up with new equipment all the time. My work was never finished, and I constantly had to drive my engineers very hard to stretch our money and keep up with the demands of the market. I was worrying constantly, not only about schedules and assignments, but also about the precision and complexity of the equipment we were designing and the incredible reliability required. We could never forget that when our equipment stopped working, it effectively shut down the whole manufacturing line of our customers, the semiconductor companies, with huge costs for them.

I lived under pressure and passed it on to the people who worked for me. Once, when I was leaning on an engineer who seemed to me to be pacing himself, he stood up, lifted his arm in the Nazi salute, and said, *"Heil, Hitler!"* to me. It was a very embarrassing scene, but it didn't really affect me because I felt the comparison was so inexact and unfair. He apologized the next day and quit his job. He didn't need to: I was not really mad at him. He was Norwegian, and I felt that his outburst probably resulted from anger he had stored up against Norway's Nazi occupiers. We have actually remained friends to this day.

I worried constantly, but everything was actually going well. Our latest mask aligner gained us a 45 percent market share, and with a price five times higher than before I had started, we were making a good profit.

There seemed, however, to be no end in sight. Each time we climbed a mountain, we saw a whole new range blocking our path. The semiconductor industry kept demanding ever-better machines with which to process bigger silicon wafers into bigger, better chips with tighter tolerances, which would produce yet more business for them, and then again for us. Late in 1971, as the recession was ending, a new Silicon Valley firm, Intel, announced its first programmable chip: this "computer on a chip," the 4004, would make feasible the development of the desktop computer. Sometime during the following year someone invented the name *microprocessor* for it.

The pressure to keep ahead of—or just to keep up with—these rapid developments was intense. As I understood business better, I started to second-guess Johnson's decisions. Sometimes I accused him of not acting quickly enough, and I often had the feeling that he considered me a pain in the butt. One day we clashed over how to handle a new competitor. Two Kasper employees, German technicians, had left to start a competing company. After I studied their product, I recommended that we buy them out in order to eliminate potential competition. At the time, it would have cost only a relatively small amount, but we did not do it because Johnson thought they would not succeed. I can still hear him say, "Forget those technicians. They will not make it."

By then Kasper Instruments had the holy sheen of financial success, respectability, and prestige. We were riding high. From where we were, it was easy to forget that Kasper Instruments itself had been started by a technician, George Kasper from Fairchild Semiconductor—probably the place where those two technicians had been recruited. We seldom thought about the company's beginnings and where our product came from—the first rough sketch on a piece of paper; its crude breadboard stage; and the first clumsy naked prototype with insufficient operational capabilities and functions. Kasper's product was now dressed in its fine, industrially engineered and designed housing. It looked so handsome and worked so well that it had found wide acceptance in the industry. Now it had that extra sheen that in olden times a good craftsman's work of art took on once it had been given to a king and placed among his treasures.

I wish I had been strong enough in my opinion to convince Johnson that the two little German technicians might make it, because they did. After they were acquired by an electronics company that helped them tune up and automate their unit, they competed head-to-head with us and eventually won a considerable market share.

Still, we were doing well, and I, too, was prospering. In 1973, tired of living in apartments, I bought my first house. I intended to bring my ailing mother from Germany so she could have the benefit of the more advanced medical care available in America. In the fifteen years since I had left East Germany, I had only been able to see her once. She had been allowed to come to Munich in the fall of 1969. We had spent three wonderful days together, and I had showed her the Alps, which she had always wanted to see.

But shortly after I bought the house, I received a letter from my brother saying my mother had passed away, almost on the day the purchase closed. By the time I got the letter, she had already been buried. This news really hurt me. Of course, I could not have attended the funeral anyway: a prison term was still waiting for me in East Germany, and, as a specialist in aligner technology, I would have been a good catch for the Russians. I was surprised how much my mother's death shocked me. I realized I had done most of the good things in my life just to please her and to repay her for her efforts in raising me. Not very long afterwards my grandmother also died, eliminating one more of my ties with Germany.

After the death of the two women who had raised me, I realized that over the years I had become completely integrated into American life. It had been years since I had even thought about returning to Germany. I enjoyed being among the many smart people in the Bay Area and being part of the cutting edge of progress in the Silicon Valley culture, where new high-tech companies were springing up daily and growing like mushrooms.

In 1974 a whole new generation of aligner products were developed by our competitors. Their new aligner technologies, which used large projection lenses or mirrors, made it possible to produce more complex chips than could be made with our current proximity-and-contact systems. Developing projection systems required huge amounts of capital. Seeing that a major market in aligners was developing, a few very large companies, including Perkin-Elmer and Canon, had bought into the game. Working closely with chip-buying and chip-making companies like IBM, Texas Instruments, and Intel, these giants were ready to invest tens of millions of dollars with the expectation that they could dominate the market. I realized then that the stakes had been raised in the game of developing the next generation of chip-making equipment—it was going to be very difficult for underfunded players like Kasper to stay in the game.

Our only hope for keeping our lead was to work harder and harder. Every day of my week now had sixteen working hours. After working ten or so hours at Kasper, I would go home and work six more hours there, thinking things through: figuring out how to build existing gear faster and cheaper, inventing new features for our machines, calculating costs, evaluating workers' performances, and assessing the likelihood of reaching our goals. I usually ended my day at midnight, falling into a deep sleep until three in the morning, when I would wake up paralyzed with fear that the competition might invent something new that could knock us out of the market. Unable to go back to sleep, I would write down all the ideas I could think of, take a stiff drink, and sleep until the alarm clock rang at six o'clock. I would then look at my three a.m. notes, have some coffee, and go back to work. Some nights I did not sleep at all, but spent the whole night trying to invent something necessary to make our equipment work better. When I was at work, I rarely went to lunch with others. I preferred to eat by myself and to use the time to revise and add to my priority list, or "shit list" as I called it. For some strange reason, it never got shorter.

Almost before I knew it, six years had been consumed in working for Kasper. During that stretch I had no social life to speak of and lost contact with all the cultural aspects of life. I never even had time to furnish my house: for almost a year after buying it I slept in a sleeping bag. Eventually I bought some cheap garden furniture and put it in the living room, but my possessions could have fit into two suitcases. I could have packed in an hour and been ready to go at any time. I had become a part in the great machine of Silicon Valley, laboring to give birth to the Information Age. I was designing and developing the tools for it and suffering the pains—like so many other people caught up in this process.

At Kasper our resources were stretched to the limit. All of us had to wear many hats. We could not afford to hire more people, and, if we had been able to, I doubt whether Johnson would have been willing to delegate authority to any of them. It became obvious that Kasper's growth might be limited not only by our inadequate resources but perhaps also by Johnson's system of centralized control and power, of having to have his hand in everything.

And yet I understood why Johnson worked the way he did. To begin with, he was a perfectionist. Also, he actually was able to do many jobs better than most of the people he had working for him or could have hired. Later, when I was in similar positions I copied many aspects of his management style—and sometimes found myself in the same quagmire.

At this time, Johnson decided to develop a second product line consisting of wire-and-die bonders—the machines that bonded microchips to ceramic substrates. He placed his bet on a very capable man who would be my equivalent in the bonder field. Soon this new undertaking siphoned off money badly needed for new aligner equipment. Worse, the man directing this project quit, and I had to take on the bonder line, in addition to what I was already doing.

Getting as many orders as we needed was always a

challenge, and became even harder during the frequent downturns that affected this fast-growing industry; during these, companies had little money with which to buy capital equipment. Sometimes getting our payment for the equipment we had delivered required another struggle. Too often companies considered delaying payments to suppliers such as Kasper a great way to stretch out money. Once, I remember, we shipped equipment to a major semiconductor company "on trial." For more than a year they used the equipment in their manufacturing line without paying us for it. When we finally forced the issue of payment, they told us they would be willing to buy it and pay—but only if we replaced it with our latest machines.

During these tough periods, Johnson cut all kinds of corners to save money. Once at a trade show in New York, he made the sales and service personnel all sleep jammed into one big hotel room. Any savings that might have been achieved through this arrangement were, however, wiped out when, at three o'clock in the morning, with everybody asleep, a clever burglar took advantage of this unusual communal situation to grab handfuls of loot; when someone awakened, he fled.

Some of these economizing policies may have furthered the company's interests, but they were hard on the employees involved. At trade shows, those assigned to duty were not allowed to sit down—and booth duty lasted all day and into the night. At one show in Munich, when the booth schedule allowed us no time to eat lunch, our Director of Marketing and Sales and I sneaked off to a concession stand and bought some packaged salami slices, which we ate hastily on our way back to the Kasper booth. Unfortunately, the salami was spoiled. We ended up with severe food poisoning and spent the afternoon kneeling at the toilets in the public restroom, while cleaning ladies scrubbed the floor around us. During this unavoidable absence we might have missed some important customers: Johnson was very upset. Severe stomach pains kept us from sleeping all that night. When we came to the show next morning, both of us looked completely green in the face; we told Johnson we were too sick to work. But he would not let us go. An hour later, while I was explaining the aligner to an important customer, everything in my stomach came up. I had to put both hands over my mouth; and while mumbling, "Excuse me," I fled back to the toilet.

We all were wondering how much longer we could survive this pace and pressure when, in 1975, a big company unexpectedly expressed an interest in buying Kasper Instruments. Cutler-Hammer, based in Cleveland, Ohio, was at that time a manufacturer of low-tech electrical devices such as switches and relays. They saw Kasper as a vehicle to get them into high-tech products, enhancing their image in the investment community. Here was what I had been waiting for: an end to the race, a way out. Fortunately, our board of directors approved the offer, which was based on a tax-free exchange of Kasper shares for Cutler-Hammer shares. As a result, we became part of a large company that had a very good future. Before the deal was closed, Johnson and I had to sign a "not-to-compete" agreement, stipulating that we would not accept employment with a competitor or start a new company in the aligner field. However, as they did not offer me an employment contract, I was free to leave the company whenever I wanted.

Cutler-Hammer's buyout gave me enough money to let me feel I would never need to work again. I realized that I had reached the goal I had set for myself six years ago when I had taken stock of my life at Hewlett-Packard and had quit.

After those years of working long hours under continuous pressure of responsibility, I felt as tired as a marathon runner who has reached the finish line. I knew I had to do something else for a while.

I had always felt that my education had been deficient, and I had always wanted to learn French: as a start on a new life, I decided to go to Monterey to take a five-week immersion course in French. (I was able to keep on learning afterwards until I became able to enjoy the treasures of French literature in their original form.) After the course, I spent ten days with Terry on the Tahitian islands of Moorea and Bora Bora and had one of the best times of my life.

When I called Bob Johnson to tell him I was back, he stunned me by saying, "It would be best if you stayed at home for a while." I was flabbergasted and could not figure out what this remark meant: was he trying out Ed Phillips as my replacement? Ed and I were good friends; he had become the backbone of all engineering at Kasper, but I could not imagine him doing my job as well as I did, and getting along with the all-controlling Johnson on a daily basis might be difficult for him. Since I could not figure out what was happening or how Johnson had meant his remark, I decided after a while to ignore Johnson's directive. I went back to Kasper, and no one ever said anything more about my return.

Even with new money flowing from Cutler-Hammer into Kasper, things were not easy. Johnson now had to live with somebody looking over his shoulder. He looked very pale and gray, and was developing a bad heart condition. I saw him taking huge pills every day. Having become a millionaire did not seem to be enough to make him happy.

After three months, I decided that this was a good time to end my industrial cycle. I would do something new, or just goof off. I wanted to walk in forests and on beaches again, do some reading, and rediscover what life was about. I handed in my resignation.

21

Suddenly I found myself with no more schedules, no more priority list, and complete freedom to do anything I wanted. I spent some time around my house and garden and some time in the Santa Cruz Mountains. In 1968, after my return from South America, I had acquired several acres of land there with a creek and a grove of virgin redwood trees, some of which were over 2,000 years old, three yards in diameter, and 200 feet high. In 1969 and 1970 I spent fifty Sundays building a cabin. The property had no electricity at that time; like a pioneer, I worked by myself, using hand tools for most of the construction work. I did use a chain saw to cut lumber from some very old redwood beams left by one of the former owners. He had brought them from the old Los Gatos Inn after it had partially burned down. Jack London had often stayed at that inn and had written parts of his books there. When I was fourteen years old, I had loved to read London's stories, so those old beams held a special significance for me.

I also painted some pictures, started to write a book, spent some nights in discotheques and bars, took friends out for lunches and dinners, and visited some money managers to learn how to make money grow. I found it hard to relax in the Bay Area, though, because there weren't many people who could spend time with me. All the people I knew were busy with their jobs during the day. They could not even play tennis with me at night; their families needed them then.

While I felt pleased at having accomplished one of my chief objectives in life, making all the money I could need, I soon realized that I had not made any plans for what to do after achieving this success. I did not know what I wanted to do next. Now that I had shifted my transmission into neutral, all the psychological debris accumulated during my life came to the surface of my mind, preventing me from thinking clearly. My life seemed to have become a black-and-white movie. I hoped that if I traveled to see other places, I might also be able to bring some color back into it and perhaps be able to see myself more clearly. Then I could shape my future. Not having any better idea, I decided to travel westward, retracing in reverse my trip around the world. I had enough money that this time I could stay in good hotels and eat well enough to avoid getting sick.

A week later I checked into the Moana Hotel in Honolulu. The next morning I woke to the sound of the gentle waves washing the shore. I stepped out on the balcony and looked over the wide expanse of the glistening, blue-green Pacific under a cloudless sky. The beauty of the scene relaxed and soothed me and made me glad I had come.

I had intended to spend only a few days in Hawaii, but I found that I was enjoying eating my breakfast on the terrace, reading the newspaper at leisure as I had never been able to before, drinking coffee served by the friendly hotel staff, observing the show below of people coming and going, walking the beaches and judging every girl's merits on a scale of one to ten, meandering through the halls and malls, and watching people flirt and dance in discos at night. Each time I jumped into the ocean, I felt the hard protective coating I had built up during the years at Kasper softening and the weight I was carrying easing away, until, two weeks later, when I jumped into the water and felt free and weightless, I began to yell, "I feel no pain! I feel no pain! I feel no pain!"—making everybody around me wonder whether I was crazy, high on drugs, or just taking a new medication.

I met some gorgeous young girls during my first days there. When I went out with them one night, I noticed how much older I had become: dancing all night to deafening music and watching them blow and snap their bubble gum during the day left me dissatisfied.

I could not get myself to continue on that trip around the world. After some time I realized that my heart was not in the project. It would be lonely, and doing it with money would take the old fun out of it.

Undecided about what to do next, I kept walking along the water's edge, so deep in thought that the bodies on the beach melted together and became a homogeneous background. I started to evaluate all the money-making ideas I had ever had in my life. Which ones had had the most potential? Which were easiest to carry out? I still regretted having given up on my slide projector, and I scolded myself for having practically given away the particle counter and having never pursued the smoke detector idea.

I kept thinking about the DNA scheme, the most intriguing idea I had ever had. It offered the possibility of continuing life after death. The idea was to create a service that would deep-freeze small samples of clients' DNA for future cloning. The enterprise required only minimal physical capital—an old building, some good refrigeration equipment, and lots of small vials. Starting the business would be inexpensive, and clients could be charged a moderate fee. One could send subscribers a small clipper, enabling them to take the samples themselves. Only three or four employees would be needed: one person would take care of the samples, an accountant would keep records, and one doctor with a good and honest reputation would answer questions and attend medical meetings and panel discussions.

The service would include storing anything a person wanted to take with him into his second life—books, personal recordings, pictures, and other articles. These might make it possible to awaken later to one's own previous consciousness; much knowledge, and perhaps even one's soul, could be preserved.

A number of possible extensions of this enterprise occurred to me. For example, we could also easily preserve samples of other life forms, especially ones from endangered species. My fascination with the incredible programming power contained in a tiny sample of human DNA led me to imagine breakthroughs of the most astounding kind: with so many different samples of DNA double spirals stored, perhaps one could design a machine to analyze their information content. Could we find out how every living thing evolved, duplicated, and built itself? Beyond that, would it be possible to discover how all different species had developed, or even to discover a secret early instruction code that may have allowed living matter to decide how to adapt itself to changes in the environment, to grow wings or a hundred feet or the carnival costumes that some species wear to attract females or to camouflage themselves?

When I had confided this idea to several friends, they had rated it anywhere from "very ingenious and workable" to "bizarre." But I had never found an honest doctor farsighted enough and brave enough to stake his reputation and career on it in exchange for a possible fortune.

As I continued walking the beach, I told myself, "You don't have to do anything. You have proven yourself already, and if you try anything new, you could fail and lose your fine reputation and all the money you just made." My mind would not consent to stay idle, however. I began thinking about Silicon Valley and where the semiconductor companies were heading. It seemed clear that the market would keep demanding more and more complex devices, including microprocessors with more memory and more functions on one small silicon chip. New equipment and perhaps new techniques would be required to create finer and finer structures on silicon wafers.

The existing technology, which I had helped develop, used lenses and mirrors in the photo-lithographic process to create the fine structures on the devices. Many engineers believed that this technique had been developed to the limit of what was possible, and that an entirely new technique would be needed to produce the next generation of devices. Two new approaches were believed to have the most promise—electron beam writing and x-ray lithography.

I realized the semiconductor industry was reaching a crossroads where it was likely to choose the wrong path. I just didn't believe that anyone could come up with practical equipment based on x-ray technology. And E-beam writers had an inherent technical limitation: to write finer and finer lines, one needed ever-smaller writing "points"; but the smaller the writing point, the longer the time that was required to write the lines in an area. It also seemed likely that the cost of an E-beam writing machine would be $2 to $3 million dollars. Such an enormous leap from the $200,000 cost of the newest, best photo-optical machines would require chip-making companies to greatly increase their capital investment. Perhaps some companies could manage to find the capital, but I couldn't see how such expensive E-beam writers could ever do the work fast enough to make it possible to sell the devices they produced at a price as low as the market would demand.

I began imagining another path. If I could improve the photo-optical wafer stepper I had invented and built in 1969, the resulting machine could align and expose finer structures more reliably than it now did, and help produce future-generation devices that would cost less than those produced by other machines. I began to imagine a much more sophisticated wafer stepper. My new equipment would have to be able to meet the industry's new demands by being able to print lines on devices only a fifth the width of the lines that could now be made, and by aligning the fine structures with ten times greater precision, to better than 0.1 micrometer—about 1/500th the width of a human hair. It would have to operate automatically, and at a high speed. The machine would have to achieve a resolution of pattern detail, an overlay accuracy, and a throughput speed which were all but unimaginable at that time. I began to see approaches that might achieve these goals. As I started to visualize the details of the machine, I estimated that it could sell for about $200,000—about as much as the best existing aligners, but only a tenth of what an E-beam machine would cost.

I knew I was the carrier of all the ideas and most of the technology needed for such a device, and, with my background, was perhaps the best person to execute the idea. I had a vision that comes only once or twice in a lifetime. I knew I had a clear shot at building a new company, being one of the few seed crystals in the semiconductor industry soup on which an important, massive new crystal might form.

Several elements of this enterprise particularly appealed to me. I was excited by its difficulty: this project was nearly as complex as going to the moon. While the semiconductor industry was always working at the edge of the impossible, what I envisioned would increase the difficulty by at least one order of magnitude. The machine would have to maneuver the silicon wafer on which multiple devices would be made with hairline precision at split-second speeds. Each of the many steps—such actions as laterally displacing the wafer 6 to 10 mm to its next exposure site, precisely focusing on the wafer surface, aligning the wafer with the reticle, and exposing it—would have to take less than one-fifth of a second. In addition, inasmuch as I was already financially secure, I could treat this challenge like a sporting event. At forty-six, I was still young enough to gamble a few years on something like this. If I won, my life's monument—my professional pyramid—would rise quite a bit higher. I might even become a hero or a demigod like Bob Johnson. Finally, I would be able to launch the project on very little money. I would not have to decide whether to risk everything until after I had enough information with which to judge the chances of success.

I kept walking the beaches, refining my idea and thinking how I would go about making it real. I worked at this idea with as much intensity as if I were once again a refugee from East Germany hungry for money and success. I don't know whether it was the balmy air or the Hawaiian sun shining on my head, but I had laid an egg and gone into a brooding cycle; like a bird, I could not leave the egg alone anymore. I was hooked. In the next few days, I shifted almost unintentionally from neutral into first gear to overcome my own inertia, and began to think out how to form a corporation, hire the right people, find money, and set the new venture in motion.

Hardly daring to believe in what I had envisioned, I flew back home and talked Dave Hopkins, Cutler-Hammer's Vice-President of Corporate Development, into releasing me from my not-to-compete agreement regarding wafer steppers. The truth was, I think, Cutler-Hammer only gave me a release because everybody thought this area was insignificant, unlikely to be important for the industry in the future; by staying on friendly terms with me, they thought they could count on my help when it was needed for the Kasper equipment.

Once this agreement was reached, I called Ed Phillips, told him about my idea, and said that in the beginning I would finance everything myself. Without much hesitation, Ed said, "I want to be part of this!"

I bought some desks and drafting boards, cleaned out a bedroom and a study in my house, came up with the name "Optimetrix" for the company, formed a corporation, and had business cards printed. My new high-tech adventure began officially in May, 1976. Ed Phillips joined me on August 2, and soon after we hired several other engineers to join us.

We began working day and night. My house turned into a workshop for engineering enthusiasts. Terry and her black cat *Abend* (Evening) were living with me then. Soon Abend became paranoid about the disturbances and started to lose all his fur; the veterinarian had to put him on tranquilizers. Terry could not take it, either. She objected to living twenty-four hours a day with a bunch of engineers and decided to move out.

I missed Terry a lot, and felt I was giving up something valuable. She had always wanted us to get married, but I knew I would have no time to spare for love and quarrels about marriage, and I could not give up the opportunity Optimetrix presented. Not much later I lost Terry completely. Someone else came into her life, and she got married.

Now the point was approaching where I would have to bet all my money on this venture. Doing that scared me. My mind yo-yoed up and down about whether this idea was sound. But we were going strong, turning out drawing after drawing: as Ed's electrical eraser rubbed more and more graphite off the drawings while he made corrections, my white carpets began turning into dark gray ones.

Even if I put in everything I had, this project would eventually require more money than I could possibly come up with. I thought the time had come to seek outside financing for the project. I wrote a business plan and called Tom Perkins at the venture capital firm of Kleiner,

Perkins, Caufield, and Byers. I was on a friendly basis with Tom, whom I had met on my second day in California when I had applied for the job at Optics Technology, where he was in charge of marketing. We had met again several times when we both worked at Hewlett-Packard. Recently I had sought his advice on investing the money I had made in the Cutler-Hammer/Kasper merger. I thought he was my best bet for getting early financing.

Tom asked, "What's up?"

I said, "Tom, I have started a new high-tech company and am looking for financing."

When he asked, "What kind of a product is it?" his voice sounded as if he doubted that I could do anything much.

I thought I deserved more respect on the basis of my track record. Now I felt unsure, and only said, rather vaguely, "It is an instrument."

His next words came out ice-cold, as if from a grave. "What kind of an instrument?"

I replied, "I cannot really tell you yet," and finished up the conversation with some polite phrases, ending with "I'll be back in touch with you when I am further along."

During this short conversation I realized how difficult and complex raising money privately would be, and that it was too early to do so. To get financing, one had to have a successful track record, which I had, and also a product one could demonstrate or talk about freely, which we did not yet have. Without that, one had to go to unprofessional investors who were willing to buy a pig in a poke—lunatics, really—and one had to pay dearly for that seed money by giving up large chunks of the company's stock right off the bat.

We continued designing and inventing, and I continued paying all the bills and salaries out of my own pocket. About three months into the project I received a surprise call from Dave Hopkins. He told me he had made a mistake, and gave me notice that Cutler-Hammer was withdrawing the release they had given me in writing. Because they had faxed the release to me, it probably was not legally binding on them. I was now in a twilight zone, not knowing whether I could or could not legally compete in the aligner field, and in steppers in particular. After only three months of life, Optimetrix faced a legal battle that could ruin me. With Cutler-Hammer's resources and all its corporate attorneys on one side, and me with my shrinking reserves and my one lawyer on the other, I knew I had little chance of surviving.

There was nothing to do but continue working while I negotiated with Cutler-Hammer. I looked everywhere for a solution. Although I knew that as an employee of Cutler-Hammer, Bob Johnson had to side with his employer, I visited him a few times to feel him out and ask his advice in the matter. I never was able to determine, however, whether he realized what the true market potential of steppers might be. I had another motive for talking with Johnson, too. I knew his talents as a corporate executive. Should it became necessary for me to give all my time to inventing and developing our product, I thought I might ask him to run Optimetrix.

That possibility disappeared soon afterward because Johnson's health deteriorated and he had to have a heart bypass operation. During his recovery he developed hepatitis from a blood transfusion. Worse, while he was so ill, the government filed a case against him. While I was still employed at Kasper, the FBI had suddenly come in one day and gone through all the company files. I had never understood what they had been looking for; now the government accused Johnson of having sold and shipped some Kasper equipment behind the Iron Curtain. Still recuperating, Johnson did not have the strength to fight the accusation. He pled "no contest" and was sentenced to five years in prison; fortunately, he was put on probation. This case ended his career at Cutler-Hammer and left him untouchable for me as well. I never learned whether he had kept all these dealings secret to protect me, or was afraid I would stop him.

Subsequently, he lost a great deal of his money in bad real estate deals. Some years later, he had to file for bankruptcy. I had always trusted Bob Johnson's judgment and his Stanford MBA degree, but I learned the hard way that his skills did not necessarily transfer to real estate. I had invested along with him, and I lost a large portion of the money I had made with Kasper Instruments.

Some months later Cutler-Hammer and I worked out a solution: they would reimburse me for all expenses already incurred, and would finance us in the form of unsecured loans; in exchange they would receive a percentage of the company and an option to buy Optimetrix several years down the road at a price based on sales and profits generated. I felt I had almost no alternative, but one reason I agreed to the deal was that the Cutler-Hammer executives seemed to me to be fair people. I was sure I could work with them, since the deal they had drawn up had the potential to benefit both sides. Also, it would free a large portion of my assets to be invested in several other start-up ventures, giving me the security of diversifying my investments and creating a better statistical probability of growing the money at a good rate, and, with luck, even multiplying it.

Now Optimetrix had the funds to go into high gear. We moved out of my house into a new facility, hired a core team of about twenty-five bright, capable engineers and support people, and worked in great secrecy, telling outsiders our project was a new sewing machine. We kept reshaping our project as we discovered more about the needs of the market and about how the machine would actually work. And as we learned more about the machine's application, it kept getting more complex. We had now concluded that each machine would have to sell for $750,000. I often thought we were crazy to attempt such a huge undertaking. Any sensible businessman would have told us it was far beyond what a small, underfinanced team could hope to do. Even in our own company, for long stretches only very few of us believed we could succeed. Nevertheless, everybody stayed excited about the project and worked enthusiastically and with great determination.

After we had increased the scale of the project, I could see we needed more money. I went back to Cutler-Hammer to sell them the new situation. They, being very understanding, provided an extra million dollars in loans without changing the repurchase formula or demanding a larger share of the ownership. This convinced me that going with Cutler-Hammer had been the right choice; I was dealing with smart, fair-minded people.

Our project was an engineer's dream. We were developing the most interesting and complex machine anybody could imagine. What kept us working was the incredible thrill that came when we had a great idea, or when we tried something very difficult, and it worked! We were so caught up in the excitement of making breakthroughs that for years almost no one could tear himself away to take a vacation. It was like designing our own big erector set—something I had always longed for when I was a kid. Only this task was unimaginably more complicated, requiring us to work at the edge of what was possible in a number of different technical areas.

The machine's optics in particular required us to push the limits of existing technology., To be able to print the extremely fine windows onto the wafer and to have a big enough area to be photographed to cover a big microchip at once, our optics had to have about ten times more resolving power than the biggest telescope on earth. Fortunately I managed to persuade the people at Tropel, an optics company located near Rochester, New York, to help us. They were the best in this business in the United States. With the help of big IBM computers, they designed our optics according to our specifications, and built and tested them with the utmost care and precision.

Because our machine would carry out a series of complex operations at split-second intervals, it would have to be controlled by a computer. We first thought we could modify a Hewlett-Packard hand-held calculator to control the functions, but we soon realized we needed something much more powerful. Our solution was to use Intel's latest CPU as the core of our computer. What we were doing seemed to me like building a larger version of HAL, the mission-controlling computer in the movie *2001,* and then preparing massive amounts of software with which to program our HAL so it could direct and sequence the operations of the many valves, motors, solenoids, stages, indicators, sensors, television cameras, optics, focusing devices, shutters, handling systems, shock absorbers, and other gadgets in the machine.[1]

If our HAL had had a consciousness allowing him to understand what he was being asked to do, he almost certainly would have shut off his circuits and played dead. Being a machine, though, he was a perfect stoic, immune to fear, doubt, and fatigue: he controlled the many functions diligently, so that the air-bearing tracks floated the wafers to the stage platform, and the stage moved swiftly to any particular location out of billions of addresses where the wafer would be aligned and exposed; he then stepped the stage to the next location for the next alignment and exposure, and so onward for each subsequent repetition; and he did all this with incredible speed and accuracy. He never forgot to focus the optics to the wafer surface before alignment and exposure. He never failed to juggle the whole machine on pillows of air to eliminate any vibration coming from the floor. He knew at any one time where in the complex operation he was and what the next step was going to be. He did not know that the stage weighed 50 pounds; he just commanded it to accelerate, then slow down extremely quickly, then stop after a 10 mm distance within less than one-twentieth of a second, without any ringing or vibration, dead on the button in a precise location no bigger than a human hair split five hundred times. Our HAL never forgot he had to finish and complete the alignment and exposure operation in one fifth of a second. Having no consciousness, he was never anxious about his performance—but we certainly were: just one malfunction by a mechanical or electronic part or one little slip by HAL, and the entire erector set might have crashed, taking with it Optimetrix and all of us who worked there.

With our limited financing, we knew that we had to solve all these problems quickly and develop a machine that prospective customers could watch doing its job. Only then would we have the credibility to sell machines. The constraints of time and money put us under tremendous pressure. At the point where we began to see that most of our problems could be solved, I made a very risky decision, gambling the whole company. We would skip the usual bread-board and pre-production prototype stages, and go into production right off the drawings, building six units, each of which we estimated would cost us somewhere between $350,000 and $500,000. We could not project the cost precisely because we were still designing portions of the machine and improving other portions. We really had no alternative, though. We had to reach, as quickly as possible, the point at which we could demonstrate, sell, and deliver units, because we would need prepayments from those sales to finance production of these and additional machines.

It was now 1978. In two years Optimetrix had grown to be a company of more than eighty employees. Our finances were stretched to the limit, but our very tightly interconnected team worked wonders as we put together our first six units. The production floor now looked more like a machine factory, with fabricating and testing stations lined up side by side, and people everywhere assembling and testing parts.

At this critical juncture, our fate took an unexpected turn. Cutler-Hammer was taken over in an unfriendly tender offer by the much bigger Eaton Corporation, a Midwestern company specializing in the manufacture of automobile parts. That business worked on the basis of developing a product, making it in large quantities, constantly seeking tiny economies that would shave fractions of a fraction of a penny off the manufacturing cost, and pumping out that product for years and years. As a result, Eaton had a huge corporate staff of economy-minded managers and lawyers. The gap between this Midwestern company's perspectives and those of a Silicon Valley entrepreneurial company was, as I was to learn, unbridgeable.

Soon all the Cutler-Hammer executives disappeared. Even the perceptive, understanding Dave Hopkins was cut loose, replaced as our overseer by an Eaton vice president. There was obviously a very large mismatch between the Eaton business unit he managed, which was used to dealing with entities producing billions of dollars in sales, and an East German refugee's ridiculously small start-up company. That manager did not have much time for us, and we always had the feeling that he was waiting until Optimetrix ran out of money. Nobody at Eaton had put his reputation on the line by financing Optimetrix, and no one there committed himself to make this inherited investment pay off. We entered a kind of corporate cold war in which the pioneer spirit that favored taking risks was wrestling against the much more pragmatic attitudes of an established company with obligations to the public financial world.

I visited Eaton's Cleveland, Ohio, headquarters several times: I never could convince anybody there that they had a great deal with us, and that continuing to finance Optimetrix as Cutler-Hammer had, could bring them a larger profit than they could obtain from investing their money elsewhere. While they could see the potential for Optimetrix to become a business bringing in several hundred million dollars a year, they seemed reluctant to invest more money and share the returns with us.

Once I met the Eaton vice president who controlled our fate in his hotel room in Monterey, California. Because we had to hurry to get to another meeting, he received me in his underwear, apologizing. While putting on his pants, he mentioned casually that Eaton was going to exercise its option to buy Optimetrix early. Since we had sold nothing as yet, they would get the company for nothing.

He knew we were completely out of money and very vulnerable. Soon after that meeting we received Eaton's notice that they were exercising their option. It would have been easy to throw in the towel at that point and walk away. Fortunately, our contract with Cutler-Hammer gave me the option of delaying Eaton's takeover for another two years, and I exercised that option. Over the past few years I had carefully built up and nursed a good relationship with a big Japanese trading company. I traveled to Tokyo and worked out a deal by which the company would take five of our units, with a pre-payment due at the signing of the contract. I came back to California with a check for $1.5 million.

We now could show prospective customers a working machine. It was exactly what the industry needed at that moment. After we demonstrated it to National Semiconductor, we received an order for twelve units, with one-third of the sales price forwarded as pre-payment, one-third to be paid on completion and before shipment, and the final one-third to be paid on acceptance in their factory. We were elated and flying high. I felt as if I were skiing through fresh champagne powder.

Eaton's executives were quite surprised. Unable to believe our reports, they demanded to see the contract with the Japanese and the purchase order from National, and we showed those. They were impressed but hardly enthusiastic.

We kept building and shipping units, hiring more people to build them as we improved and refined our product. Our growth was limited by the amount of cash available, however. Meanwhile, established companies in the mask-stepper field modified their units to make them able to align and photograph images on wafers. Our customers knew these established companies, and could be confident they would survive. Buying from us meant taking two big risks. If Optimetrix went out of business, the customer would be stuck with an unsupported machine. And if our new equipment stopped working, our customers' fabrication lines, which cost them $100 million or more, would be idle until our units were fixed. Any long stoppage would cause heavy losses. We needed enough capital to keep up with the more established companies and to exhibit a level of solidity that would reassure our customers. But in spite of all the reasons for giving us the capital we needed to compete, Eaton continued to stand by, contributing nothing, waiting for the moment when they could take us over.

Early in the spring of 1981, I met Eaton's CEO and Chairman, Mr. DeWindt. We had a straightforward, eye-to-eye conversation about the situation in the semiconductor equipment business, the extent of my ambitions for Optimetrix, and Eaton's fear of making mistakes in a field that was new for them. They wanted to expand into attractive high-tech ventures to improve their image and pump up their stock, but they knew this strategy involved serious risks for the corporation. They were in a bind with us: if they gave us more money, we would build up Optimetrix and Eaton would then have to pay a higher price for it. On the other hand, if they did not give us money and we did not expand, they would end up getting a company which had lost its edge, and their previous investment would be wasted.

I understood DeWindt and liked him, and I think he liked me, too, different as we were. We had one basic thing in common: both of us had started at the bottom, he as a stockroom clerk at Eaton and I as a mechanic's apprentice at Carl Zeiss. As our meeting ended, he said, "I will think things over and see whether we cannot move now."

A week later he called to say, "I am sending a deputation of twelve people out to look at Optimetrix. Please accommodate them."

After my "Okay and thank you," I told him I hoped we could work something out. But I did not have a dozen executives to match his: I was on my own. For moral support I asked a friend, an entrepreneur running his own company, to join me. I only remember one moment of that meeting—a line spoken by my friend, who was supposed to help me. We were on the production floor watching our people assembling and checking machines. Suddenly my friend pointed at me and said, "Without this man, this assembly hall will be a junkyard in five years!"

I could have killed him. This was not an idea I wanted Eaton to have. As part of my deal, I wanted to be free. Fortunately, though, the Eaton executives ignored the remark and in fact did not seem to recognize what was making Optimetrix so great. They recognized the excellence of the product and the potential of the market, but they were unfamiliar with the dynamics of Silicon Valley, where a high-tech company could only maintain a lead by providing an assembly of top-flight engineers with challenges that would spur their creativity and with incentives that would maintain their enthusiasm. In the semiconductor industry, even a revolutionary new product could be obsolescent by the time it was first being shipped. If the process of inventing, designing, and producing ever more sophisticated equipment did not continue without interruption, the competition would devour your company.

A few days later Mr. DeWindt called me and said, "I have thought things through and am prepared to make you an offer, take it or leave it. But first I want you to understand that I have discussed this with my executive committee and my board of directors, and this is a very tough decision for us. Eaton is offering to buy all the outstanding shares that you and the other employees of Optimetrix own." When he named the price, I was stunned. His offer was about one-quarter of what I thought our shares were worth.

I said, "But . . ."

He cut me off, "Think it over for three days. This offer is set in stone."

[1] Later, when our successors wanted to alter the programming, thirty software engineers worked at the task for three years without achieving their goals.

22

DeWindt's offer seemed too ridiculous to consider seriously. Still, I decided to take two days off to think things over. I drove to Sea Ranch, a quiet resort one hundred miles north of San Francisco. At this beautiful place one can walk along the shoreline or the bluffs above and enjoy unspoiled vistas of the ocean, the broad meadows, and the redwood-covered hills.

In assessing Eaton's offer, I thought about everything I had done since I returned from Hawaii to form Optimetrix. I had succeeded in building a stepper that was now helping to make much more sophisticated new-generation computer chips. While running Optimetrix, I had learned on the job how every part of a company's complex mechanism and procedures works; how a company lives off ideas and inventions; which of these warrant investment in design, development, and testing; how the resulting products are put into production and sold; and how these products lead to ideas for further products. I had done everything I set out to do and felt proud. If DeWindt's offer had only been higher, that would have crowned the whole achievement.

In evaluating the offer, I had to consider the effect of the changes in my personal life that had come about while I was creating Optimetrix. While I had been reevaluating my life on the beach in Hawaii, I had thought about my status as a bachelor. Even though I had been looking all my life for the perfect woman, I had never found one with whom I wanted to spend my entire life. While I had sometimes come close, as with Erika, Christa, and Estella, more often I had tried in my mind to assemble one perfect specimen from desirable physical and mental attributes of several acquaintances. I had given up on this pastime, though, because assembling a virtual woman tends to twist one's mind.

In general, I could not believe how little time and effort people put into their searches for a partner, or the compromises most of them eventually made. To me, selecting a partner for marriage seemed to deserve as much thought as selecting and pursuing one's education or profession, yet often people married after knowing each other only a few weeks. Many of the marriages I had seen left me feeling that I should avoid altogether such an imperfect relationship. In truth, I was repelled by the lives many married people led. They spent more money than they earned and struggled with their children. Many wives worked like slaves, toiling endlessly in their kitchens and chauffeuring their kids to and from school, sports events, and other activities. Most parents spent time worrying that they were not doing enough to help their children achieve what they themselves had never accomplished. After years of marriage, partners often had little love for each other and, on the rare occasions when they had sex, had to fantasize about someone more pleasant and sexy to gain satisfaction. Many people married to have company, but after years of living together quite a lot of them seemed to be lonelier than single people like me.

On the other hand, it was obvious to me that men, consciously or unconsciously, did all their achieving and posturing for women, and women all their achieving and beautifying for men. Even at the age of forty-seven, I loved women and looked at them as works of art. It was important to me to be taken by a woman's beauty, to see and feel warm skin and textures, to stare into naive eyes and down into the soul, and in the process lose myself as I had when I was eighteen. What disturbed me was that one could not complete a relationship as one can a painting, and then start a new one. A marriage seemed to remain always a work in progress.

I had never been particularly interested by the idea of having or raising children. If one really wanted to have kids, I thought, one could always find a woman who wanted to stay independent, and work out an arrangement by which she would have one's child and one would take care of the woman and the child financially. One could even set up several such arrangements. Carrying out such a plan would remove the blind chance most people took in having offspring, allowing a more controlled and rational selection process. It would also allow one to stay free, continuing to focus intensely on challenging projects, pursue esoteric ideas, or live on the fringes of society among artists while still allowing one to see one's sprouts growing whenever one wanted. One would begin in the state in which a huge percentage of marriages ended, with the woman caring for the children and the man paying for their support; but in this case, the children would not experience the desertion of a divorce, and the parents would avoid a great deal of frustration and anger.

When I thought of the way I was living, it seemed to me that I was trying to keep both sides of my nature in balance: whenever the hard-working, successful industrialist took on an additional commitment or responsibility, I had to add a freedom of the same weight to the bohemian who needed to feel able to wander freely and to follow the impulses of curiosity and the desire to create. I sometimes fantasized that I could avoid marriage, stay young, and live my life's unusual equation by getting young, bright graduate students to fall in love with me, the bohemian artist; then, as each graduated and found a job elsewhere, I would look for her successor. I knew that breakups were painful for both parties, but at least such relationships would not get stale.

I never analyzed exactly why I felt that the women I loved had to be young, but the few times I had gone out with older women convinced me that I did not have the interest or the patience to listen to all their problems with children, ex-husbands, and boyfriends, and take on a share of the responsibilities they had accumulated. I also had a tendency to picture in my mind how they would look ten or twenty years down the road—which often frightened me. I knew it was stupid to think that way; if I operated on my plan of not staying with anyone longer than a year, what difference did it make how someone might look in twenty more years? But I could not help it; these images and feelings were always there.

I never thought of myself as old or older. I always surrounded myself with young people. By absorbing their energy and thinking, I felt and acted like them rather than like people my age. As I became ever older, however, I could not ignore that more and more young women looked away when they saw me or looked right through me, as if I did not exist. Like a good businessman, I then tried to reduce my demands, settling for my loving them, ignoring their not loving me; but I never felt comfortable as a sugar-daddy. So, despite the ever-lengthening odds, I returned to the ultimate challenge of finding a perfect young woman who also loved me.

But life has its ways of providing opportunities to those who seek them. One day one of my younger friends invited me to play tennis with two young ladies he had just met. When I arrived at the tennis courts, I found him sitting with them in the Jacuzzi, warming up for our tennis match. I was immediately taken by the serene beauty of one of the girls. Her name was Irene, and she looked about eighteen. Her face and shoulders reminded me of the finest, most delicate porcelain, and when I talked to her, I had a feeling of looking into a calming, well-pruned Japanese garden. I could hardly play tennis and was glad when we quit. When we were alone for a moment, I said quickly, "You are a wonderful person. I would very much like to get better acquainted with you. I need you in my life."

Not quite believing me, she laughed and replied, "Okay, just call me sometime."

I did not forget Irene, but during this time we were selling Kasper Instruments, and then I was set to go to Hawaii and then around the world. When, instead, I returned to start Optimetrix, I needed a secretary. I talked to several applicants without finding one I liked. I called Irene to see whether she would be interested. She was, and we met for lunch. I found out that she was twenty-four years old and had held six secretarial jobs. I told her what she would have to do, even though I concluded from our conversation that she could not follow instructions or respect her superiors, and perhaps could not even type very well. But she was very beautiful, knew all about the celebrities from reading newspapers and magazines, and could talk very well about any social subject. She was so real and such a natural delight that I spent more than three hours with her. I kept asking more and more questions about how she did her job until I inadvertently asked, "Can you cook?"

She looked somewhat astounded and said, "No, I cannot, but I can learn. Why?"

A few days later, we played tennis together, and eventually I invited her to spend a weekend in Monterey with me. She told me, "I have never spent a weekend with a man, but I'd like to come with you."

I was not at all sure how this trip with such a young woman would develop and added, as if I were a doctor, "But I am always on call and might have to return at any time."

Irene was Eurasian, and I sensed her Japanese mother had taken great care to protect her unspoiled nature. She was as sensitive as a mimosa and as gentle-natured as a dove, but she could lash out at you if you attacked her or did something she thought wrong.

We drove down to Carmel-by-the-Sea, checked into the Highlands Inn, and walked along the beautiful coast talking about lots of things and our lives and views. In the evening Irene surprised me by showing up in my room for a nightcap wearing a beautiful negligee. A little embarrassed, she told me she had bought it especially for this occasion. The next morning, while we were eating our breakfast and talking, she paused for a few seconds, and, looking straight into my eyes, said suddenly, "Heinz, I am going to marry you."

It sounded to me quite naive, cute, and very unrealistic. I just laughed and said, "Really?"

I don't know what happened to me, but on our next date I suggested that she should go back to the university and finish her degree, staying in my house during that time if she agreed to take care of it. She accepted, moved in, and once in a while even cooked a meal for me.

Once, when she had served dinner with potatoes that had gotten cold, I made the mistake of commenting, "Irene, I like my potatoes served warm."

She broke out in tears, got up, and saying, "How dare you?", packed all her things in her Volkswagen and drove off.

After a few days I found her and talked her into moving back in, but at the slightest negative comment she would clean out her closet, walk down the stairs with all her clothes over her arm, pass through my living room where I was pretending to watch TV without seeing anything, and, without a word, throw everything into her car and drive off. In time her drives got shorter and shorter, until she just sat awhile in her car in the driveway. Later she carried her clothes only halfway down the staircase, and finally left them permanently in the closet.

In 1979 while we were walking by a jewelry store in San Francisco, I made the mistake of buying her a small emerald ring she had fallen in love with. As soon as I gave it to her, she said naively, "Now we are engaged, right?"

Thinking to myself, "Never do that again," I told her, "No, Irene, this is just a friendship ring."

But Irene lived from then on as if we were engaged, and I could not do anything about it. By that time I was the only person she seemed to have in the world; after she moved in with me, her mother and father would not speak to her.

In 1980 Irene graduated. For me, this was supposed to be the time to change partners, but Irene came to me and said, "What should I do, Heinz? I know I should get a job, but I would also like to have a child."

Here was the chance I had been waiting for, the time to make an arrangement and set up my first child-rearing unit. I found, though, that I did not have enough guts to propose this idea in the real world. I had, of course, always realized it was only a thought experiment. On the other hand, I wanted to avoid complications now and needed more time to think things through. Since I could not find anything wrong with Irene—she was as perfect as one woman could be—I decided to leave things alone and let her stay, recommending that she read all my books and make herself useful. I began thinking seriously about her future and designing a good one for her. I decided that the best thing for her would be to leave me, get married, and have children; however, my mind kept shifting back to an encounter I had had just a few weeks earlier, while skiing in Courchevelle, France. While I was riding up the chair lift with a German from Hamburg, he told me, "The best thing I did in my life was to have kids. I wish I had more." This had surprised me and had sent me off thinking.

Once I had traveled to Porto Alegre, Brazil, to explore a business opportunity. As I sat in the hotel lobby, waiting for my room to be cleaned, I observed several Brazilian upper-class wedding parties going to church across the street, and, after the ceremonies, parading by me through the lobby for receptions in the hotel. I was struck by how young all the participants were, a reflection of the fact that the Brazilian population was, on the average, much younger than that of the United States or Europe. What really affected me viscerally was how incredibly vital and natural all the young women and men looked. I sensed in their beautiful skins and animated expressions all the young, healthy blood running through their bodies. Seeing them made even my own blood more liquid and tempted me to join in and become part of their world, a world that was of the future even more than of the present. Absorbed in this scene, I could feel for the first time in my life how life's chain was extended by young people like these, vowing to live together in love, and to care for the new life they would generate together, the next links in the chain. That night I thought back on those visions of life's continuity. As a bachelor I suddenly felt, in some basic sense, useless. No occasional adventure with a woman could stand up to the incredible joy, harmony, unity, devotion, and purpose I had just witnessed in these newlywed couples.

Perhaps, I thought, a man like me who wanted to do everything exciting in life was missing something fundamental—having children was the only opportunity for the continuation of life known to man. I bounced these thoughts around in my mind and let them ferment for several months. Then one Friday morning in November of 1980, I heard myself saying to Irene, "This is going to be a beautiful, sunny weekend. Why don't you pack a few things; we can drive to San Francisco and spend the night at the Stanford Court Hotel, and maybe we can get married tomorrow morning?"

From me, Irene found this suggestion surreal. Even though she thought I was joking, she packed a small suitcase, in which she included a wedding dress she had purchased, unbeknownst to me, just in case.

From the hotel I called Muir Woods National Monument. I said, "Excuse me, when does the park open in the morning?"

Mr. Hershey, a ranger, told me the park opened at nine o'clock in the morning.

"Excuse me, Mr. Hershey," I said, "is there a chance you would show up at eight and let us in? I want to get married at the United Nations Redwood Grove before the park opens to the public. Also, I was wondering if you would be so kind as to be our witness?"

Hershey thought I was playing a joke on him and kept laughing about this crazy idea, but when he finally decided I might be serious, he said he would be there, let us in, and, if necessary, be our witness. I then called several ministers. Some did not believe me, others did not want to get out of bed so early. Finally I found one who needed the money. He asked why I wanted to get married at Muir Woods. I told him that even though I wasn't a churchgoer, I believed in marriage as a Christian ritual and thought of the two-thousand-year-old redwoods as in some way symbolic of the beginning of Christianity, and therefore a perfect setting for our wedding. This convinced him that I was serious, and he agreed to come. I also called several photographers, but none were willing to get out of bed so early on a Saturday just to shoot a wedding picture.

Next morning we drove to the park and met the very agreeable Ranger Hershey and the minister. As we walked beside the gently murmuring creek under the giant redwood trees, a great blue heron led us on, flying always a few yards ahead of us, settling down as if waiting for us, and then taking off again. It was as serene a scene as one could imagine, a setting out of a fable, and Irene looked like a gentle dream-elf just hatched into this world. In the United Nations Grove the minister married us and gave us his blessing, as did Ranger Hershey. I was really moved by the ceremony, but thought it a bit nonsensical that the pronouncement included the phrase "for as long as you both shall live"; but then Irene turned to me and said, "For me, this is forever," and in kissing me dispelled the doubts that lingered in my mind.

After the ceremony we had a wonderful brunch at the Madrona Inn in Sausalito, overlooking the San Francisco Bay. Later we drove to the Golden Gate Bridge and walked to the middle, where we hugged and kissed again. With our best wishes, Irene threw her flower bouquet into the water below us, and we watched it float slowly out into the Pacific Ocean. We had the luck to find a photographer at the other end of the bridge, ready to capture this happy moment for us. Years later, it still seems one of the happiest of my life.

Now I had to make another decision of great consequence—what to do about DeWindt's offer. Not only was the offer too low, it also would require me to give up my self-made identity of "Mr. CEO and Chairman" and the power that was mine as the head of a successful, essentially independent business entity. All of this was difficult to part with. I had come far enough to know that I did not need these titles to be myself, and after twenty years of hard work, I was tired of river-rafting down the Class V rapids of the Silicon Valley industrial stream toward a future of unending further rapids.

As I drove home, I painted in my mind the scenarios that would result from my decision on the Eaton offer. If I rejected it, Optimetrix would continue to lack cash. In business, nothing stands still: in spite of our success now, we had to start developing a second generation of steppers, and after that a third, and so on. I knew that the major reason young companies failed in Silicon Valley was that, as product cycles grew shorter and shorter, it became harder and harder to generate enough cash and to retain or obtain the top talent needed to upgrade the old equipment and design the new. If the economy took a downturn, we could find ourselves with so little liquidity and credit so stretched that when a customer canceled an order or delayed a payment, we would be unable to meet payroll.

If I accepted Eaton's offer, I could greatly reduce my own financial risk. With the resources of a large company behind them, the jobs of Optimetrix's employees would be more secure. In fact, in many ways Optimetrix and I were already married to Eaton. Whether I wanted to or not, it seemed I had to yield and accept the compromise.

If I accepted the offer, the capital from the merger would provide a new and much larger nest egg from which I might be able to hatch myself as a small, independent venture capitalist and financier. I could use what I had learned launching and building companies to help people like me start up interesting new enterprises. If I put half of the capital into several risky but promising situations and served on a few boards of directors, I might be able to create an interesting life for myself. I would have a chance to make more money and also have time for a parallel career as a painter; perhaps I would write as well, and read all the books I had always wanted to read. I would be able to play sports regularly to stay in shape, travel a lot, and have fun. Even if I lost all the money, it would not kill me. Having lost everything several times before, I knew I had the stomach to accept losses.

When I got home, however, I hesitated. I told Irene that I was going to call DeWindt the next day to tell him I had not been able to make up my mind yet, and let him hang over the cliff-edge of uncertainty awhile. Perhaps he might break down and make a higher offer.

I honestly had little hope of that happening. Eaton approached business in a careful, pragmatic way. Their salaried executives were used to protecting themselves and to minimizing failure. They were quite aware how much risk Eaton was taking to get into the high tech area.

Irene listened very carefully, nodding her head to each point as I spoke. She did not seem quite herself, and I thought for a moment I was talking over her head or was not making myself understood. That wasn't it at all. She took a deep breath and, almost apologizing, said, "Heinz, there is something I have to tell you, and I don't know whether this is a good time, but I visited my doctor today and he told me I am pregnant."

I felt for a second as if I had just been drafted into another life with responsibilities I could never shake off and bonds that could never be untied. But stumbling toward Irene and taking her into my arms, I asked her, "Are you happy about it?"

Irene did not answer at first, and we hugged in silence for a while. Then she asked me, "Do you want it?"

Not believing myself, I said, as if I were getting married again, "Yes, I do!"

The next morning when I talked through my thoughts on the Eaton offer with Ed Phillips, he said, "We could invest that money in something else. We can always start up another company at any time. What I don't think we can do is fight the empire much longer, but it's really up to you."

For me, Irene's condition had changed the equation. The next day, with a heavy heart but with no uncertainty, I picked up the phone to call DeWindt. "Hi," I said, "I know I'm three days late, but I have decided to take you up on your offer."

DeWindt said only, "Are you sure?"

And I replied, "Yes."

"So, at this moment you and I have a handshake on this deal?"

I again said, "Yes."

He said, "I will set the wheels in motion."

And I said, "Go ahead."

This was a deal between two men. It was all said and done. From this moment on I felt equal to this big guy from the empire.

But in reality, our handshake was only like the conception of a child. Eaton and I had to live through a nine-month pregnancy to complete the merger. Lawyers wrote a contract that grew into a pile of papers nine inches thick. Attending to its thousands of complicated questions and decisions absorbed all the time I could steal from running Optimetrix. For nine months, my one lawyer and I battled the army of Eaton lawyers and Eaton's corporate administrative mechanics.

Even in corporate affairs nature runs its course. Shortly after my son Klaus was born, so was the Eaton-Optimetrix merger. New money from Eaton began flowing into Optimetrix operations just in time, and all of us were relieved. We had made it to a safe haven.

23

The payment from Eaton's buy-out did not come in any dramatic form, such as stacks of gold coins or hundred-dollar bills. It came as prosaic Eaton stock certificates, pieces of paper that looked like birth certificates. Holding them in my hand, I thought back to the time in East Germany when I had earned my living working ten hours a day, bound to the semi-automatic revolver lathe like a human robot, breathing oily air, manipulating levers, belts, and gears, and jamming my tool bits into metal to curl chips away from the work piece, earning only fractions of pennies for each usable part, making just enough money to live another day. I had promised myself then never to forget those times and never to spend money frivolously.

When I handed Eaton certificates to employees in exchange for their Optimetrix shares, I told them that Eaton had given us more than they had to; if they had kept starving us of cash and waited, they could have acquired us for almost nothing. I also told them that Eaton's stock had a very good chance of appreciating and should be kept. Even at its current value, it represented a considerable stake for us all—for some, a year's salary; for others, more like ten years' salary; and for me, enough to build myself a new future.

I remember not long afterwards walking with Irene on a beach at La Jolla, near San Diego, jumping up and down and singing deliriously, "I have made it!" I had cried out in triumph in the same way when I graduated as an engineer. This time I was far away from Jena, realizing the American dream: I had a family, a house, money, and freedom. More than that, I was now conscious that I had already lived a good, interesting, fruitful, and complete life. I had enjoyed almost all of it and, most important, it had never been boring.

I felt that whatever days of life remained to me were extra—icing on the cake. I even hoped there would still be some ups and downs, perhaps even a few harmless nightmares, to make things interesting. When I looked back now, I realized how important even the bad periods had been. One has to live through such bad times the way one gets through scary movies like *Psycho,* gripping the seat until it is over, knowing that at some point the terror will end and one will step out into the sunlight.

I was satisfied, and in some ways glad, to have been born poor. My situation had forced me to think about my goals and to work tenaciously and take risks to attain them.

All things considered, I was even content to have grown up without a father. Though I had often missed him, his absence had allowed me to select my own father figures and mentors as I grew toward adulthood. His early death had also prevented me from seeing him perhaps fail as an artist, grow old and frail, and die. He had always remained for me as young and non-threatening as when I sat on his lap at age four, marveling at this person who could create things and do magic.

Even though I had always intended to sell Optimetrix someday, the actual closing of the deal left me feeling somewhat uncertain about what I should do in the near term. While I was free to leave, bound only by a new not-to-compete agreement, I found I could not just walk out the door and say farewell to Optimetrix's affairs. Indeed, I was still Optimetrix's president. With Eaton's approval I had already hired a man who would be the new manager of the operation. As he learned, I handed over control to him as fast as I could. During this period DeWindt and I got along very well, Optimetrix thrived, and Eaton wanted me to remain. But after two years I was ready to go. Optimetrix was then absorbed into another Eaton semiconductor division—which had grown out of Kasper Instruments.

While I was phasing myself out of Optimetrix's affairs, I thought a great deal about what I should do with my future. How would I keep myself from running so hard? Should I spend my time painting, gardening, reading books and studying philosophy? or maybe live as a ski bum or a wind surfer? or try to climb Mount Everest? Should I buy a better house in the United States, or Hawaii, or Germany? Should I just sit at home with my family and raise my kid? or leave everything and start a new life—give all my money to charity or a religious sect and become a hermit on a mountaintop? or start again from scratch with a new venture?

Once I really was free, Irene and I talked for days about all these thoughts. Recognizing that I now found myself in uncharted territory, anxious about the future, she kept her cool and just said, "You have worked hard all your life to reach this point, and you should do anything you really want to. Do only things that amuse you. I will go along and never cage you in this house. Klaus will not need you for perhaps three years; I can take care of him by myself for a while. If you want, you can take that time to go out and explore possibilities and choose a new life for all of us. I will always be here and will wait for you, no matter what.

"Our money is now safe in Eaton stock, but eventually you will sell those certificates and have to manage cash assets. That will not be easy. Money has a life of its own and will take up a lot of your time. You might even lose all of it because managing money is not your profession; but if that happens, we will only be back where we were before. That possibility doesn't bother me, because I have read a lot in my magazines about how the rich and famous live, and most of them are not all that happy. We are not quite as wealthy as they, but you can go and see whether you would want to live the way they do. Just make sure you stay who you are and don't get so confused that you end up with your feet in two different boats. Some rich people want to be in two places at the same time, but of course they cannot; and so they are never happy in any place, no matter how wonderful it is. Remember, life is not like TV; it has only one channel."

Irene told me that she had just read a great article about Zermatt, where people ski right under the Matterhorn. "That is something for you. Why don't you start your new life with a ski trip to Switzerland? I saw pictures of a great hotel where wealthy people stay—it's called Mont Servan. I am sure you will have a good time there, and maybe you can come up with something good for all of us."

I was not sure at all about leaving Irene behind, but she insisted I should go by myself. Since I couldn't talk her into coming along, I finally agreed that maybe it was better for both of us to have some time for contemplation. I knew I was certainly ready for such a trip. I spent a few days buying a tuxedo and dress shoes, new ski equipment and boots, and a first-class ticket to Geneva, and then spent another few days in my small library, polishing up my German and French by reading some of my favorite books aloud. I also felt I had to leaf through stacks of other books to prepare myself somewhat for the Europeans. I was always somewhat apprehensive and insecure when I encountered European intellectuals because they loved to talk about their knowledge of the inherited—and in my opinion anachronistic—cultural load they had acquired during their humanistic education. While I knew my short review course would not get me far in a game of "intellectual chicken," it at least made me feel better.

Irene took me to the airport. When I kissed her goodbye, it was with great love, but the departure really seemed no different than leaving for one of my many business trips.

After the plane took off, I stretched out in my first-class seat to read the European newspapers, with interruptions from the flight attendants serving drinks, then serving food, then trying to induce me to buy perfumes and other duty-free items. The movie began, and I started watching it, distracted by defective earphones that distorted the sound. I had to pay so much attention to understand any of the dialogue in this boring movie that I got tired of listening and dozed off several times. Finally I took out the earphones and watched the movie without the sound. Moments later, I slipped into a half-sleep that left me unable to tell whether I was dreaming or still trying to follow the movie.

I seemed to see myself on the screen, walking down a road into what looked like a huge fog bank, quite like the ones that come off the Pacific Ocean and roll over the mountains near my house. I just kept walking through this haze until I came to a very large open office area, where thousands of clerks sat at desks. They were helping the hordes of people who seemed to flow in from every direction. When I got closer, I thought I recognized one of the clerks. He saluted me. It was Smilo von Lütwitz, my brother's *fahnlein führer* (young folk leader), from 1942, but his face was scarred, like that of a young man I remembered who had been severely wounded at the front. He had volunteered to go back and fight again. I knew that he and Smilo had died in battle a few weeks later, and so I was confused at seeing Smilo sitting before me with the other man's scars.

But when I came a little closer, Smilo was not there anymore. I saw instead the grinning face of my interrogator from the East German secret police. He was in uniform, and I thought immediately about the ashtrays he had thrown in my face, and remembered what he had said when he closed out my case: "Even if you escape now, or if you ever get out of prison, there will be a time when we will come where you are, and you will be damned. Nobody can stop Bolshevism." I had never forgotten his words, and had often had nightmares about them. I knew immediately I had made a big mistake coming to this place.

Before I could say anything, he also disappeared, turning into my Culligan man. Shortly after I had moved into my house in Los Altos, this salesman had rung the bell, lifted his uniformed hat, and said, "I am your Culligan man." I had been so impressed by his uniform that I had stood at attention listening to him, and instantly bought a five-hundred-dollar water softener on the strength of his word that if I didn't buy it, the pipes of my house might clog up with calcium tomorrow. I had often thought about that episode as an example of how my German past could still influence me strongly, even after fifteen years in the United States, but I had not expected to see this man again. He still wore the same Culligan hat, lifted it as before, and waved me over to his desk. I tried to click my heels, but then realized I was barefoot. Noticing this, he smiled and said, "Even though you are German, you don't have to do that kind of thing here."

I mumbled, "I am sorry."

He smiled again and said, "There is no reason to apologize. I guess you have become a real American now. You have our habit of apologizing for nothing."

I said, "I am ..."

He interrupted, " I know who you are. I have seen you coming from afar, and this is not the first time I have seen you on your way up here. Once again, though, it's not your time yet, and I must turn you back."

"What do you mean?"

"Well," he replied, pointing to a large, bright flat card on his desk, "we have this computerized cardex system that keeps track of everyone on earth, and we can easily distinguish between those who are really dying and those who only think they are."

"Really?" I said, "I can't believe this, or that you are here."

He looked at me and said, "I remember you particularly well because you were one of my last sales. I died that same day in a freak accident in San Francisco when a crane collapsed."

"I'm very sor ... " I said, cutting myself off, asking instead, "Why is this place so big?"

"Well," he said, "think about it. At the rate people are dying down there, we have to handle 300,000 new arrivals every ordinary day. Then there are another 500,000 or so like you, who think their time has come. We have to process you false alarms quickly and send you back. That's a lot of work. Without this incredibly well-organized electronic cardex, we could never manage."

Intrigued by all this, I tried to charm him into telling me more about his cardex system and how it worked. It seemed that he still liked to explain technical equipment: he pushed toward me the big flat card. I saw it was a very thin flat-panel TV screen. He told me these cardexes indexed and cross-referenced all lives, allowing the staff to know every life pretty well at a glance—even its future, for their supercomputers could now use probability and prior patterning to calculate each person's life experience some fifteen years into the future. He demonstrated, explaining, "I just have to go onto my cardex, speak or type your name, and in no time your lifeplot appears. I can then tell you everything about your time on earth, both past and future. All of our admissions officers here have access to that information for anybody down there at any time."

I was fascinated by this advanced technology and wanted to ask more questions about it. He was so enthusiastic and excited I could not stop him from talking, however. He told me that while he and his organization were far ahead of our technology, they were somewhat worried by the rapid pace of information technology development on earth. "What would happen if cardexes like this were in the hands of people without the wisdom to handle them?" he asked. "What will happen when just about anyone can call up every bit of information mankind has collected over thousands of years, or examine lifeplots with every detail of anybody's past, present, and future, or instantaneously communicate with anyone else on earth, no matter where the two of them happen to be?" He doubted that having so much information and being so much better connected would make people happier; rather, he felt it likely to cause people to live under more stress and be even lonelier, as the result of having less and less direct contact with other human beings.

He worried, he told me, that when it became possible for humans to calculate each person's future, most would not be able to tolerate hearing the full story of what fate held for them, and would react by doing even more foolish things. He also feared that those who were unable or unwilling to process complex information quickly would fall behind and not be able to find employment, while the quick, capable people who could work with all that data would earn unimaginable sums of money by reducing the risk factors in any kind of undertaking. "Fortunately, though" he said, "nobody in the universe can alter destiny, read anyone's mind, or control thoughts."

I found everything he said fascinating and logical, but I realized that, like most people, I really wanted to talk about myself. What lay ahead for me? I finally worked up enough courage to ask him to let me take a look at my own file. He surprised me by saying, "Oh, sure, no problem. In principle, anybody who is going back may have access to everything we have. You won't remember it, because everything is erased from your mind as soon as you leave this office area. Just tell me quickly what you want to know, and I will call it up. We can play a few short videotapes of your life if you want—even some precalculated videos showing your future. While we've been talking, I have already started to look some things up, and I will tell you a few things you must be anxious to know—how long you're going to live, for example."

I was amazed and anxious. "Let's see," he continued. "Well, I have to take that back—we don't yet have enough information to calculate your date of death. But I can tell you that you are not coming back here for at least another fifteen years. I see you at that time still living with your wife Irene and your two sons, with all of you in generally good health."

Learning of that second son left me a bit stunned. Noticing my reaction, he continued, "Looking at your plots I see that there was some chance that your bachelor attitudes might be so strong that you would not successfully adapt to marriage, but it seems that you are adjusting surprisingly well to family life—maybe because you were unusually old when you married. Perhaps that's why you will live harmoniously in marriage in spite of continuing your adventures by climbing mountains, heli-skiing, and taking financial risks, all of which will scare the living daylights out of Irene."

I felt some momentary guilt about these activities—even though I had not yet done them!—but I was eager to look over his shoulder to learn more. I could see the cardex screen; unfortunately, without my reading glasses I couldn't read the print. I begged him to call up future financial deals. He did; as he read them out to me, some of them sounded disastrous, but others seemed to be rewarding. Nothing indicated that I would return to total poverty. I was relieved at that.

He moved to another sector—a video that showed me five years in the future, in 1987, behind the Iron Curtain in Prague for a secret meeting with my brother. I saw myself on the cardex screen, waking up in the early morning. I am rubbing my eyes, walking nude to the hotel window and looking down on a big square below. A lone man who could be my brother is walking toward my hotel, but after twenty-nine years, I cannot be sure. Suddenly I am sure it is Frieder; I know him by the way he walks. I start yelling and banging against the sealed windowpane; but I cannot open it, and he cannot hear me.

Next on the cardex screen, I saw myself walking with Frieder through Prague. I am telling him everything about my life since we parted in Berlin. He says little. I sense that he is not used to talking freely; he is used to putting everything through the double filter of real and official talk, as the party in the East has forced him to do for the past forty years. From what he says, I sense that his circumstances have not changed much since I left East Germany. He has raised two children and has taught sports and mathematics to three generations of students. His children now have children of their own. I am two generations behind and don't know any of them.

I noticed on the screen that in Prague I am frequently turning and looking behind us, as if suspecting that somebody is following me. My brother spends the night in his car; disturbed by bad dreams, I wake up at three o'clock in the morning, go to his car, and tell him to eat all the West German chocolate I gave him and to get rid of the silver foil and packaging in case he gets stopped and checked on the border on his way home. I imagine that the Stasi will accuse him of being a spy, as they did me: even chocolate wrappers might be evidence. I tell him that whenever I talk or even think about these experiences, I always have nightmares about getting apprehended, thrown into prison, tortured, and sentenced.

I was fascinated and frightened by that film segment. I promised myself to remember it and under no circumstances to ever fly to Prague. At this point the Culligan man cut off the film, saying, "I see this upsets you too much."

He keyed in a few commands. "I see you wrote many letters to your friend Blumenstein but never got any response. They were, in fact, all confiscated by the Communist authorities, but you will be visiting Jena after the Berlin Wall comes down and will find his sister, Jutta. Unfortunately, Karl-Heinz will have died by that time."

Again, I could hardly believe what the Culligan man was saying so casually—me, back in Jena! the Berlin wall down! I did not know what to ask about next, but I managed to mumble, almost to myself, "What did Blumenstein end up doing?"

His answer surprised me as much as did learning that the Berlin Wall would come down: "Blumenstein could have really used your help, because he fell into disfavor with the authorities. After that, he worked for many years as a concierge at the gate of a factory."

I tried to speak, but found I could not.

When I was finally able to regain my voice, I asked him a question that had begun to bother me. "Everything in your system seems to be based on predetermination. What about one's own will, one's thinking, one's ambition? Aren't we at least partially able to choose what we do? Can't we at least influence some outcomes?"

The Culligan man chuckled. "Well, you are making the same basic mistake in your thinking that most people do. You are missing our celestial catch-22: your thinking and all the choices you make are also part of the closed loop of predetermination. They are just as predetermined as are the people from whom you inherited your genes and the circumstances into which you were born. Nobody gets to do anything with a second brain or a second body. Nobody gets to live in a different world. As long as you live, you are always limited to the same basic equipment with which you were born, and you remain forever connected to the circumstances in which you began your life. Consider your garden: if you watch a flower grow from a seed, you can see how predetermined life really is. The same process always yields the same species of flower, and it in turn produces the same seed. Once a rose, always a rose; once a daisy, always a daisy. Of course, an individual flower may grow bigger or smaller, depending on weather and soil, and over the course of hundreds of thousands of years the species may get hit by strange particles or cosmic rays and mutate into something else, but even the timing and trajectory of those rays and particles is predetermined."

I was fascinated and wanted him to elaborate on these ideas. I wanted him to help me understand why I had always felt that I was living a surreal life, always practicing and training myself for real battles to come as I had once trained for war with my wooden bazooka. But instead I addressed him in a tone edged with some indignation, "So all my efforts and accomplishments are really not attributable to my own efforts, to my 'self'? Do you mean that 'I,' what I am or feel I am, my conscience that has instigated everything and has always called me to account, can take no credit for my successes or blame for my failures or misdeeds?"

He began to fidget, the way an important executive to whom you are trying to sell something does when he wants you to know you've used enough of his time. He replied, "Even though I have explained all of this to you as if it were the only way things logically could be, it took me a long time to accept all this myself. Your conscience is just part of the mechanism in you that executes the necessities and givens of your existence. It is more or less a natural manifestation of all the complexities of your brain. We don't seem to encounter this phenomenon yet in animals or plants, but even that could happen one day."

I wondered whether, if I had just known all this, I might have lived more comfortably, more fatalistically, with my conscience and fears attenuated. Then I remembered what he had just pointed out, that nobody can change his basic equipment and the circumstances surrounding him.

Even though I knew we were running out of time, I dared to ask one more big question: "What about life after death? How do things continue after people have registered here and enter the big gate?" For a moment he lost his competent Culligan salesman composure. But he caught himself quickly, and, leaning forward, whispered, raising his eyebrows, "I don't know. Nobody in this office knows, because no one who went through the big gate has ever returned. It is a big mystery."

He leaned back in his chair and thought a bit. Then he added, "Strangely, each of the millions of persons who came through here found the journey as natural as another trip for business or pleasure, or just to the supermarket. No one has ever remembered his own death. So fear of it seems unwarranted. All we have been, have experienced, and have done might only be a practice run, something like training with the wooden bazooka for a new sequence of life to come, and so it may go on to infinity."

I sensed this was all he dared to say on this subject. Regaining his poise, he shifted again through several visuals on his cardex screen and then said, "Oddly enough, I see you actually will change your life markedly. You are ending your industrial career prematurely and going into an artistic mode—you are painting and reading and writing a lot. The cause is not at all clear."

That made me remember one of the recurring dreams I'd had about my father, in which, already much older than he had lived to be, I put my own life aside and took up his at the age at which he had died, living as a bohemian and painter, as he would have.

When I mentioned this to the Culligan man, his face lit up as if he had found the solution to a puzzle. "That explains everything. The person that I am seeing here on the screen looks more like a painter than a businessman, yes, and that really could be you having become your father. That may also explain why, even fifteen years from now, you keep behaving as if you were a young man. Look here. At this airport, you are loaded down with suitcases and bags, but you're still one of the few people who are running even on the escalators and people-movers. And here you are heli-skiing in British Columbia at the ridiculous age of sixty-eight. I see you riding in a helicopter and then skiing down a slope in the Monashees called "Steep and Deep," going through the trees at a horrendous speed, completely enveloped in clouds of stirred-up snow. This really looks quite dangerous."

He shifted the screen once more, stared at it, closed it down, and started to sweat. Lifting his cap, he wiped his bald head, mumbling, "Quite alarming. Such risks!"

He cheered up and spoke to me again. "But of course, we know that everything is predetermined. You can get away with it, you are fixed in the 'pre-de-loop' and are a prime example of how hard it is for some people to get killed. Some people may fall into a crack, but nothing ever seems to kill them."

24

Fifteen years have passed since I returned to California after my near-death at Zermatt. Did that experience lead me to change my personality and my way of life? Or were the changes of the next few years ones I would have made anyway in response to the change in responsibilities that came with giving up control of my corporation?

Even though I remained President and CEO of Optimetrix, going to work daily for two more years after Eaton bought us out, the new conditions removed a great deal of weight from my shoulders. With Eaton's financial resources behind us we could afford to do what we needed to do when we needed to do it. Also, I was no longer the only person responsible for protecting the livelihoods of our employees.

The sale did, however, create two new problems for me: deciding what I wanted to do next, and finding ways to invest and protect my assets. Meanwhile, another new responsibility slowly came to absorb a great deal of my time and my interest: I was a father.

When my first child was born, I saw him immediately as an individual and not as part of me. I had always thought of individuals within a family as being coupled together like the links of a chain. While Klaus was unquestionably my son—he had even been born with a tiny hole in his earlobe, just as I had—I sensed nothing of a genetic connection to him. There was little in common between this baby and me, and I felt no particular attachment or gladness that my first child was born and that I now had a son. Perhaps I felt that way because of the age difference of fifty-two years, or because I had never thought of having a family and raising children. But from then on, he was always there.

Curiously, doctors calculated the time of conception exactly to our wedding night. During the pregnancy, Irene suffered from extreme toxemia, a condition characterized by elevated blood pressure and water retention. It added sixty pounds to her one-hundred-ten-pound frame. Neither of us liked the process of pregnancy, with its changes in Irene's appearance and the way she felt. Sometimes I felt as I had while waiting for American tanks to come down the road to meet my bazooka—that perhaps I should run away before it was too late. When I took Irene to the hospital, to leave her in the care of professionals who earned their living by looking after pregnant women and delivering babies, I was somewhat relieved. This way, I would delay reality and perhaps avoid the hand-to-hand combat of the final hour. Naturally, I visited Irene every day and brought her flowers as any good husband would. But I did not want to know how the baby in the belly was; my visits were just with Irene. I carefully tried to avoid thinking about the baby in there: this was hers, she had wanted it, I had given it to her reluctantly as a present, and I would monitor "the project" from some distance. I was not ready to face living with a third person or sharing Irene's attention and love with somebody else.

One Sunday afternoon I came to visit Irene, bringing her flowers and some candy. I had planned to go on afterwards to San Francisco to enjoy some hours in a coffee house drinking cappuccino and watching the scene of young people there. Irene seemed to be okay; I had just kissed her goodbye and was outside the door of her room, ready to leave, when the doctor intercepted me. Putting his hand on my shoulder, he said, "It would be a good idea if you planned on staying. I am about to give Irene an injection of pitocin to induce labor. Go see the nurse. She will show you where to scrub your hands and give you sterilized surgical clothes and a mask. In a case like this, one never knows whether a Caesarean may become necessary: I want you to be prepared. You can at least hold Irene's hand. I will try my best, and you might be able to help me."

I was shocked. We still had six weeks to go. Nobody had told me they might induce labor at this stage. It was not time, I was not ready, and here I was, about to lose my beautiful Sunday afternoon in the coffee house and deliver a baby instead. It took me more than an hour to move from my surreal coffee house mind-set to the brutal reality that two lives were at stake, and that, at best, if everything went well, I was going to be a father. I did not have any idea what that really meant, and I could not imagine myself in that role. I did not want to witness this outcome of beautiful activities eight months earlier and face the consequences.

But it all happened anyway. I did everything the doctor told me and stood by. Irene did not seem to suffer too much pain. We waited for several hours, and then when her contractions became more intense and closer together, something like a cone-shaped missile head appeared, then a tiny face, followed by a tiny body with real miniature hands and feet, attached by a cord to Irene. And then I heard his voice. It was a loud cry and at that moment I knew that this was it, the wake-up call that I had been trying forever to avoid. This is how fast one's life and status can change. I was a parent and would never forget this little voice, no matter how much I tried to ignore it; it would be in my mind even if I did not hear it.

When I saw my son lying in the incubator a few hours later, he was no bigger than a hand, and he had a tiny bit of poop hanging out of his rear end. I thought, "My God, I am glad this is Irene's kid. I don't have to take care of him; I am so glad I am a man and still have my freedom."

For the next two years I never touched the baby, fed him, or changed his diapers. Irene did everything. She carried the baby wherever she went, like the women in Third World countries who carry their babies on their backs with them even when they work in the fields.

As time went on, Klaus began to capture my attention by walking around, chewing on everything, and making his home in my home. He started playing under my desk with everything that had before been only mine. I became involved gradually, half by defending my things and territory and half by trying to protect him from hurting himself or drowning in the swimming pool. Everybody kept saying how cute he was and how much he looked like me, but I did not believe it. I always remembered the joke where a baby sticks out his rear end and somebody looking at that says, "He looks just like his father!" I got upset and scared one night when he came into our bedroom and peed on the floor. Another time he peed into his own toy box and then into my iron fireplace. I thought he was mentally ill and watched carefully, worrying, for signs of intelligence. Brushing aside my ignorance about children, Klaus perhaps thought, "I belong to you. There is no other man here; you must be my dad." And so, gradually over three years, he became my kid and not Irene's anymore.

Irene missed him and wanted to have him back. We settled the case while vacationing in Greece by deciding to have another child. We got it, delivered exactly nine months later to the day. I was again present when my second son, Hans, was born. I would not have wanted to miss it. He was a big baby, twice Klaus's weight, and I felt rather scared when I saw him slipping out. He looked strong. The strange thought occurred to me that he might eventually push me aside or do me in. But he turned out to be a gentle character, very loving, and, while extremely strong in mind and body, ready to share anything.

Klaus was born good-natured, too, and with unusual artistic talents. From age five on, he astonished me by collecting all kinds of parts, gadgets, and things from nature and assembling them into meaningful, unbelievably perfect sculptures. The only trouble I had with Klaus was that he sometimes treated little Hans like one of his toys and imposed his power over him. Generally, though, the two of them played together inseparably, as if they were newly married, with Klaus playing as if he were a younger boy, and Hans playing as if he were an older one.

As Klaus has grown up, it has often been interesting to observe how he has used his curiosity, his sense of humor, and his judgment to absorb and digest German history, composing his own picture from what I told him, newspaper stories, movies, what he learned in school, and his trips to Germany. Out of movie images and dialogue from *Hogan's Heroes* and *Those Magnificent Men in Their Flying Machines,* and from the pointed helmets of World War I, John Cleese's goose step in *Fawlty Towers,* the battles of World War II, the Holocaust, and observation of that comic character, his father, he built up a surrealistic picture of the German mentality which often tempted him to play the role of funny, curious, or serious German characters, creating various quips: "A German officer can sleep well even under the most difficult conditions." Or, when I ask him, "How are you going to learn to drive?": "Like a German officer learns everything, from the book of instructions, of course."

I don't know quite what to think of him, because he also formed years ago an imagined galactic empire and at times still plays a curious General Field Marshal from up there and is quick to appoint his fellow students to high-ranking galactic positions.

Something about kids that still amazes me even today is their very presence every morning. I wake up and have breakfast, and they keep me company. They don't ever seem to think of leaving me or Irene to attach themselves to another mother, father, or family. Even though I know that dogs and cats stay with their family, it seems a wonder when one experiences this with one's own children. I am sometimes in awe at how sensitive they are, and also at how quickly they forgive you, even when you make mistakes or misjudgments. Even now, years later, I don't quite understand how they know they are my kids. I have learned very important lessons with them—that I belong with them and that nothing is really mine. The greatest lesson I have learned is that I must share everything.

I came to spend a great deal of time with my family, traveled all over the world with Irene and our two little sons, and have never regretted that I changed my life. In managing situations and responsibilities with Irene and my children, I benefitted greatly from my experiences in industry and business, never having any doubts about what was normal or logical or tolerable. I managed to shoulder the responsibilities of family and children well, and never found my kids troublesome. I am forever grateful to Irene and for the conversation with that man from Hamburg on the ski lift in Courchevelle that gave me the final kick toward this change.

While I was experiencing in my personal life all the changes that occur as children grow, my business life was also changing. I stayed with Eaton/Optimetrix until 1984, but with the expectation that I was going to leave. Gradually I changed my function to monitoring decisions rather than making them myself.

During this period, and for years afterwards, I experienced what might be called "post-industrial stress syndrome," with symptoms rather similar to those I had experienced after my prison experience. I was plagued by nightmares in which I had made the wrong decision, had not recognized what needed to be done, was being beaten by the competition, was losing everything, or was having to start out from scratch again and again, rolling my huge rock like Sisyphus.

I was not the only executive who had such experiences: after I quit Kasper, my main competitor admitted that he too had lived in fear—getting up bathed in sweat at three o'clock in the morning, worried about what I might invent next, and from which direction I would next get at him. Also, when I was in Tokyo, Japan Canon's management told me that it was a great relief for them when I left Kasper. They had monitored the patent applications we filed, and they said to me, "Please stay out of the aligner field." Five years after I quit, they had taken over the aligner market. Six years after I left Optimetrix, they did the same with the stepper market.

As my role in Optimetrix diminished, I was thinking about what I would do when I left. I did not want to continue in my profession, or to battle elsewhere in direct management of a business. I was not ready to form another company and do it all again, or to continue climbing toward the top another way. Neither was I ready to become a nonentity, however. I had worked hard to become who and what I was. I was proud of my career and my title, "CEO and Chairman of the Board," and often thought and spoke from this identity. Indeed, I noticed that even in my private life I sometimes declaimed in the authoritative tones of a knowledgeable VIP—occasionally, even on subjects about which I really didn't know much.

My work had more than filled my days, and my schedule and working habits had become so much a part of my life that their energetic rhythms pushed me to keep on inventing, managing smart people, having a say in how and when things would be done, and creating the future. Doing anything different would require a big change in my habits and way of thinking. In some ways I had become the incarnation of the role I had been playing. I was not sure what was still left of my true self under the hard surface.

I thought often about this potential identity crisis, and about ways to fill the void that would exist when I left my job and my title behind. Fate handed me an unexpected and unusual opportunity that prevented me from having to deal with an abrupt transition.

Six years earlier, when we were building our first production models at Optimetrix, I had often hired consultants to carry out well-defined critical tasks. I worked especially well with two consultants from Hewlett-Packard, Sam Maslak and Rob Young. They did a marvelous job of designing and de-bugging analog circuits that drove and dampened the stage within the unit as it accelerated rapidly and stopped quickly. I was very impressed by their expertise and genius. I knew that at H-P Sam was or had been a project manager, working in the infant field of ultrasound. One day I asked him whether he had a desire to start a venture himself.

He told me, "I have been thinking about it."

I said, "If I can do it, you can do it, too. Go home and try inventing a new generation of equipment, and if you need any financing, call me."

I had no idea then that Sam had already left Hewlett Packard. Not long afterwards, Sam took out a loan on his name to support himself and started inventing in his garage.

About a year later, Sam called me and said he and Rob had exhausted their own financial resources and needed to raise money. I listened carefully to what they had to tell me about their project. The nature of the product they hoped to develop appealed to me: not only did it have great financial promise, but also it would help people directly by giving doctors a new tool for diagnosing disease. I certainly did not understand completely everything they said, but they made a lot of sense, they were addressing a good market, and they had already managed successfully a number of very complex and very difficult high-tech projects. I thought, "If they cannot pull this project off, nobody can." Even though their equipment would be quite different from what we had created at Optimetrix, my experience in assessing the risks involved in design and production and managing an infant high-tech company could be applied to their problems. They made me an offer; I accepted it, wrote them a check, and became part-owner in their undertaking.

From then on, my thoughts were often with them. I wanted to help in any way I could, and I had many meetings with Sam (most of them in a pizza parlor in Mountain View called Frankie, Johnny, and Luigi's). I became a sounding board for him, gave advice on how I would approach things, and eventually devolved all my applicable knowledge onto him.

Almost two years later, toward the end of 1981, when I was working with the lawyers on the papers for the Eaton-Optimetrix merger and worrying about what to do after the merger became complete, I began to sense that Sam was really going to come up with a saleable product. Several engineers had joined him and he was going to form a corporation. I knew he would need very substantial first-round financing to complete his product development.

After thinking long and hard about my situation, I had the idea of saving or transferring my industrial and business soul, the part of my ego I stood to lose with the Eaton buy-out, by putting an additional large investment into Sam's new company. By redirecting a substantial part of the money from Eaton, I could own a big enough part of Sam's company. I could then grow with him, maybe even without hands-on involvement, and with luck, a few years down the road, have as much status and ownership in his company as I had in Optimetrix.

At first I thought this idea was crazy, desperate, unwise, and dangerous, but gradually I convinced myself that taking a chance with Sam was the right thing to do. Of course, there was a chance I might lose my investment, which would set me back financially nearly to the point where I had started twenty years ago. If that happened, however, I believed I could still muster the physical and mental strength to begin a new venture and regain my financial security.

I think now that at the conscious level I greatly underestimated the risks and overestimated the probability of succeeding in another venture. I was still viewing the financial world at least partially through the eyes of an engineer, optimistically and unrealistically, and did not see the many possible combinations of bad luck, bad decisions, bad timing, and competition that could swallow up my investment. Somewhere deep inside, though, I must have sensed that risk, and tried to reduce it by continuing to be Sam's mentor, helping him wherever I could and keeping a close watch.

When I was deciding whether to go ahead with the investment, the risk did not seem much higher to me than having most of my assets in Eaton stock. If I cashed out of Eaton, I would have to decide where to put the resulting capital. I considered various other possibilities for investment, but all of them seemed to me to have drawbacks. My experiences with investing in real estate, oil wells, and the stock market had taught me how easily one could lose one's money once one left one's area of expertise.

I had learned that money was alive—a baby that was always there and would often wake you up at night. One had to handle and nurse it, could not leave it alone, had to become its father and learn how to take care of it. As with a human child, you the parent were the only one who could properly do that. But, just as I had resisted fatherhood, I opposed the notion of spending most of my time thinking about and managing my money. I also wanted to put it somewhere where it would increase, but I wanted that place to be somewhere where it had the likelihood of creating something with real value, beyond money.

While I was still struggling to become successful, I had become aware of how money surreptitiously affects the personalities of those who acquire it. For example, I noticed that when the newly rich were traveling and dealing with people who did not know them, they wanted to wear their new identity so visibly that each receptionist would give them a better room, each maitre d' a better table, and each waiter better service. Yet they really could not just tell each stranger, "I am a great man: I have done this and that, and I own all these assets." Many people would not believe them, and anyone who did might mug them.

It seemed to me that one advantage of investing in Sam's company was that my assets would be committed and inaccessible for many years, so that it would be as if I didn't have that money. I could postpone the changes money might produce in me. Perhaps this was crazy thinking, but there was some truth in it.

The time when the first major round of financing was needed arrived before the Eaton-Optimetrix merger had closed and I had the Eaton money in my hand. In order to get a second opinion I brought the topnotch venture capital firm of Kleiner, Perkins, Caufield and Byers into the game by calling on my friend Tom Perkins. They never told me how they arrived at their decision, but after they committed to share the financing of that round with me, I felt more comfortable and went ahead.

After the Eaton-Optimetrix deal closed, I invested, step by step, almost half of what I got from it in Sam's company. With Optimetrix I had reached my summit in high-tech. From it I could see many other peaks to climb, but I had no desire to do another start-up myself, unless fate forced me to.

During the period after we had sold Optimetrix I also financed some other ventures, trying to help some of my best former employees become entrepreneurs. Unfortunately, none of them succeeded, and all the money invested into their ventures was lost. I learned from these failures that venture financing is hard work. I also became much more careful.

Eventually, Sam's company became very successful, making it to the big board, the New York Stock Exchange. I did not have to begin the struggle all over again. Sam, who was an engineer and inventor when we met, became an excellent CEO and chairman of the board. The ultrasound equipment Sam developed is now widely used in medicine to examine babies *in utero*, internal organs (heart, liver, breasts, lungs), and other soft tissue. It was so good that it is now on permanent display at the Smithsonian Institution. I was extremely pleased to have helped build a company that made such a big contribution to public health. After fifteen years, I am still serving as a director on its Board. Sam's success has left me free to follow my instincts—painting, spending time at home, traveling, and pursuing other ventures in business.

Sometime in 1981 I realized that I also wanted a physical memento of my success, so I started looking for a better house. I found a very special place, one of a kind, but it was much too expensive, and too big a step up for me. Even so, perhaps because it was unique, I started to negotiate anyway, just for fun. I did not like the layout of the house that was on this land, so I obtained its plans and redesigned the house I had not purchased and could not afford. This process went on for nine months. When the owner came far down from his asking price and I thought I already had the deal in my pocket, a developer got wind of it and grabbed it for only $25,000 more than I had bid. Irene and I had become so attached to the idea of moving there that we were devastated when this occurred. But there was nothing I could do; it was gone.

I took the down payment money and invested it in a small company instead. Three years later, this company had grown so much that my stake in it was worth almost ten times my original investment. I had been calling the new owner of the property with offers above what he had paid for it. He never said no, and I called again and again until three years later, he came back with an offer. It was much higher than what he had paid, but I took him up on it the same day. The outcome was that my original down payment had turned into an amount that paid for my dream house, with enough left over for the reconstruction costs.

Redesigning and reconstructing the house gave me great pleasure. I realized that my desire to create open spaces unobscured by walls and doors and with as many large windows as possible was an after-effect of having been in prison. When construction began, I found that I very much enjoyed dealing with workers from the various trades. They had personalities different from those of the engineers, businessmen, and other white-collar people I had worked with. To me they appeared to be more normal, more easy-going, and more in touch with the basics of life.

As the reconstruction project went on, I found I had a great need to work in the open air, digging the ground and planting many bushes and trees to create better landscaping. The result satisfies and soothes me, even though on hindsight the project seems to have required as much effort as starting and managing a new company.

This house project relaxed me, and I found that being free to create what I wanted reawakened my artistic urges. All through my engineering career, I had never lost my longing to work as an artist and live as a bohemian, free from worries. I still wanted to have interesting discussions with people used to talking about ideas that did not have to be executed or reduced into products, as technical ideas did in industry. After my exposure to so much technical information and tech-talk, the range and freedom of more liberal discussions had become even more appealing to me. In a discussion of philosophy or art, one could say things that didn't have to be precisely defined, accounted for later, and ultimately turned into a tangible product. In such conversations, words and whole dialogues could flow more freely, might often go off into the nowhere, or could be more easily replaced with new and better ones. With that freedom, one's mind had a better chance to reach new, interesting and stimulating insights, or at least reach the pass leading to unexplored mental territory.

In the art world, one was allowed the kind of thinking that was impossible in the world of technology where everything needs structure, definition, and organization. In the arts, one was entitled to one's own taste, opinion, and viewpoints, especially when discussing newly created art, works which had not yet collected a weight of statements by authorities that limited viewers' ability to see the work without knowing what they were supposed to see.

In part because I gave up my flirtation with the guitar, I failed to realize many of these dreams. As a young man, I would never have believed that I was going to spend forty years virtually locked inside factory and office walls without nourishing my musical and artistic inclinations; but I never sang or played a musical instrument again.

When I lived in New York, I had promised myself I would give painting a real chance. I really wanted to explore what I could accomplish working with paint on canvas. A few years later in California, I bought real artist's paints and canvases and painted, more or less in an exploratory fashion, several pictures a year with a wide variety of subjects, sizes, styles, and mediums. I had varying degrees of success, mainly because of my free and experimental approach, and because I had never taken any painting lessons and had no pressure to sell anything. I did not paint painstakingly on canvas as I assumed normal painters did; instead, I painted without restraint or concern for detail—as a poet would write the first draft of a poem or a songwriter a song. My only objective was to create something pleasant to look at, something that could create feelings as poems did.

By the time Eaton bought Optimetrix, I had arrived at a point where I liked my paintings, and I actually thought I could make a living painting pictures and keep myself occupied and inspired without another regular job. But it really was impossible to switch from my technical career entirely into the arts.

Over time I had become more and more fascinated with modern painting. Visiting art galleries, especially museums of modern art, had become one of my favorite things to do. I liked many of the modern images displayed, especially the ones projecting views nature could not offer. They widened my horizons in beautiful ways, and they could evoke entirely new feelings. After many years, my visual intelligence had become sharpened; now I had become as enthusiastic about paintings as music. My new house had a big artist's studio. I started to paint more and more, often just applying paint, many times not knowing what I was going to try to paint or what I was going to end up with. I liked the freedom this offered and the colorful world one could create. It was fascinating that, using a minimal amount of liquid and color pigment, one could open up so much psychological depth and space within a small area of canvas supported by a frame. I liked the opportunity painting offered to create something that had no preassigned meaning. Painting provided a means through which one could bring philosophy, intuition, inspiration, and perception to fruition and offer views never seen before to oneself and other observers. I also liked the intuitive process of painting an image that showed different moods and impressions to the painter each moment paint was applied and an area was covered, leading him through a fluid process to images frozen in time. I liked that some of these paintings, while in reality just batches of dark and light paint, color or just black, white, or gray, smooth or textured, never left me with the feeling of having just looked at a wall with batches of paint, but rather of having looked into deep, multi-dimensional spaces, and often into the unknown and into enigma.

After I put together the first catalog of my paintings, I noticed that I had unintentionally created a visual autobiography: the paintings were states of mind and ideas that had found their way on canvas, each reflecting particular points of time and illuminating how I had progressed to where I was now. They were a compilation of incidents, emotions, and power searches that had driven me to paint them.

I always felt a connection between my painting and my work in industry. It was not that I thought about or painted mechanical or electronic parts, factories, or offices. Rather, painting enabled me to connect into a whole the separate segments of my life at a particular time. A German who immigrates to the United States is still one person whose life bridges two very different societies and attitudes toward life. Somehow influences connect and make a whole—as continents connected by the oceans make a whole earth. While my paintings had for me meaning in themselves, together they expressed all I did in life and what I was at the particular time I painted them. It seems ridiculous that I worked as hard painting a picture as I had designing an instrument or device. But this was my style, and I could not change my work habits. And to my surprise, because the eye can be an incredibly exacting judge, I found painting often as difficult, precise, and demanding as technical work.

Perhaps I really did die at Zermatt fifteen years ago and have since experienced the latest of my many rebirths. Prior to that event, most of my time and attention had gone into my business. Some of my time still goes into the inescapable chores of daily life that come with owning a house, having investments, and living in the world, but now the bulk of my time and attention go to being a husband and father, to painting, and to those activities that keep me able to escape meaningless routine and avoid the in-turning and self-restrictions of age.

Fortunately or unfortunately, I have never managed to live like a bohemian. That remained a dream. I simply was not born one and never obtained enough freedom from responsibility to become one.

25

As I write this, I am on my twelfth heli-skiing trip to British Columbia. At sixty-eight I keep asking myself, "Why do you still do this? Why commit to more of these helicopter flights to remote mountain tops, to be left to ski for your life through snow nobody has tracked before?"

My answer is that I am still unwilling to shut down any of my physical or mental systems. Skiing through the natural slalom of snow-covered rocks and tightly spaced trees, one must react in split seconds, intuitively, for there is no way to avoid the next turn or the force and pain of an unexpected jump. The constant danger demands instant responses, with no time to evaluate or plan as a Bob Johnson might. One flies twelve feet through the air, and then perhaps twists to manage a sharp turn, or sinks up to one's head in the snow pillow below, suddenly skiless and immobile, or emerges as a snowman, if the skis stay on, to take the next obstacle, using ephemeral visual cues and instinct to evade dangerous tree wells, avalanche-pregnant steep sections, hidden tree stumps, cliff edges, and holes over ice-water creeks.

All of us who do this have the best equipment—skis, boots, and all-weather gear—partly because bringing good equipment is the only preparation for heli-skiing that is of any value. One could never practice for this on normal ski slopes: it would be like practicing for war with a wooden bazooka. This is a real battle. Once the helicopter drops you off on the mountain peak, you stand alone, deep in powder never touched by a ski, peering at an extensive glacier below and at the endless forests that cover the steep slopes farther down. You then have no alternative: you must forget your fear and the severe pain your muscles signal to your brain after one hundred turns or so, and focus on the joy of flying, almost literally flying, down the slope. And you do. After a day of ten helicopter drops and skiing 30,000 elevation feet—the equivalent of skiing Mount Everest from summit to sea level—one feels so much better, younger, reconfirmed and happy that all systems are still working that one remembers only the good parts of the silent runs, or the stretches where one simply had to shout out loud in disbelief and joy.

It is a joy still to be able to ski with these groups of people much younger than I who come from all over the world to experience this. Sure, I could take it easier and ski at Lake Tahoe, or perhaps in Colorado or Europe, but heli-skiing is like mining for rare minerals: the risk-reward ratio makes it all worthwhile. For me, going heli-skiing is in a way like releasing the water that has built up behind a dam all year to surge through the turbines and generate electricity. The power of heli-skiing is electrifying.

Each time I come up here to B.C., I feel some fear and apprehension about whether the snow will be good and skiable, and whether I still can do it. Sometimes I am tempted to turn around and forget about it; but then I meet the other forty lunatics who have scheduled themselves for helicopter flights to mountain summits, to ski through whatever lies between the top and the pick-up sites more than 3,000 feet below. I think, "These guys are all too young for me; it will kill me skiing with them—all athletes full of energy, bump skiers with flexible knees, downhill racers who can never go slow, Warren Miller heroes ready to jump cliffs, spread their legs in the air or make somersaults and helicopters, capable of skiing "death crust" any time.[1] Compared to them I am just a Sunday skier.

Everybody tells me, "You look great, you are in good shape, you are skiing like a beast!" But I have seen the x-ray pictures of my teeth: only I know which ones are capped, where teeth are missing, and where the bridges are. Only I know how fearful I am, when I forget something or cannot remember a name, that Alzheimer's may be setting in. I put my head in the sand when it comes to assessing the rest of my body's bones and mechanics. What frightening pictures they might yield!

This is real battle, though, and there is no place to hide. I will be hooked up with these guys for a week, with no choice but to make the best of it. My time in prison taught me that one has to make the best of everything; one has to pull oneself through these apprehensive periods and beat down the dog inside that says, "Quit, put on your Hush Puppies, sit by the fireplace, get the wife to serve food in front of the TV, watch the news of some skier getting killed in an avalanche, and hope for a weather report announcing warm temperatures and rain in British Columbia, which will make the decision to stay home a good one."

Once up here, I get sucked into the vortex and do the best skiing I have ever done, experiencing what may be the greatest excitement available to a human. It always amazes me how quickly a group like this can blend together, how forty or so men and women of different nationalities and professions can amalgamate in no time if they have just one common objective, to ski like hell. And though I often behave like a hermit at home, for me getting together with such people is as important as the skiing itself.

Like the guys I was imprisoned with, each person here has a story—the appointed executive who runs a big company, the Mormon executive who has inherited a family-owned firm, the man who has made something from nothing after starting his own venture. A stockbroker sells the shares of such companies; a venture capitalist finances new companies; an accountant keeps track of the numbers; a physician repairs the others when they break down. A shipbroker sells huge tankers to customers from all countries. A hotel owner runs his chain of twenty hotels like a smaller Club Med. He and his partner started from scratch when they were twenty-one-year-olds with long hair. They had no staff: when businesses called, the partner played the role of each department head, changing his voice for each job.

A politician from Texas descends from ancestors who arrived there so late they could only settle land where nothing but desert rats and rattlesnakes could survive and cattle went begging for food. After generations of starving, more oil was found under their land than anywhere in the state; so much money was made so quickly that it all got lost like rain in hot sand. So the family brought back the cattle, while the politician stayed in the city to live.

One guy's great-grandfather once owned and was governor of a whole province in Mexico. After the revolution, his family lost everything in the land reform. Another, who lost $25 million in the stock market in a single day, says, "Actually, after all the money was gone, I felt relieved. It was pushing on my back to do something with it, and I could never make a decision or change my life."

There are trust fund babies, fighter pilots, brain surgeons, bridge builders, lawyers, architects, and developers. And there are, of course, those ordinary wage-earners who have saved money for years to pay for this week, to spend it all racing down virgin slopes as the culmination of their skiing lives.

I love an effervescent Irishman named Desmond who tells jokes, often jokes of the kind folks tell in pubs after everybody is drunk. He laughs at each one as if it were new, making us forget our self-made problems involving wives, families, fortunes, jobs, and failing sex lives. I love his very Irish accent, his attitude, his earthy goodness, his honest wisdom at fifty, and, even though I don't drink, his affinity for ale and whiskey.

I also love the food the chef prepares every day; he performs and acts like a star. The man who previously ran this lodge in the wilderness died last year. He was Swiss, and I sensed he was worried; he was only fifty but looked much older. Trapped by his job, he was probably worried stiff about his future: how long could he still do all this work and put up with guests who sometimes don't listen to good advice and don't follow the rules designed to prevent someone from skiing down a cliff, causing an avalanche, or endangering others? Some people are just plain reckless.

I am fascinated by an art dealer who deals in rare letters, real or perhaps forged—it is often a matter of opinion. With an attraction to life's precarious trails, he windsurfs and skis well; he lives a life among celebrities, politicians, and the wealthy, has appeared on *Larry King Live* and on *60 Minutes,* and is often in courts to testify as an expert witness in trials about signatures and documents, real or forged. He profits from people's need to become part of history or to become famous by possessing something from someone who already is or was. He might sell the letters of Winston Churchill or the diaries of Jack the Ripper. Everything seems to go well until someone asks for authentication or verification of personal letters of a famous person such as Sherlock Holmes, with no idea that fakes of a letter from a fictitious character like Holmes might one day be more valuable than an authentic letter from a real person. All it takes is a client with a good sense of humor who realizes that even fake letters can be unique. Real or fake, what is the difference, I ask myself; but I am very impressed when he sends me a book he has written about documents and signatures and their verification. Among the many documents in the book is one of Hitler's with his signature. It really intrigues me; Hitler's signature looks absolutely bizarre and bedeviled.

I find it fascinating to ski with a man who deals hands-on with history's paper trail, who can clarify who wrote what when, who could perhaps even untangle the Soviet past that I saw changed so often while I lived in East Germany that it became more unpredictable than the future.

At sixty-eight I worry more than the young about the things that may go wrong. I have seen skiing partners up here who have never once used the recommended leapfrog technique, even during a high-avalanche-danger period like this week. If caught in an avalanche, they would be without a back-up. Anyway, I am sure that in any such situation, most of these people would be incapable of finding a person and digging him out in the five short minutes of opportunity before oxygen deprivation turns him into a vegetable. Yesterday I was apprehensive about one guide who, I learned, had started just seven weeks ago. It takes time to learn this territory of five hundred square miles, with its five hundred runs through dangerous territory. As a result of this guide's lack of familiarity with the region, we got lost on our first run and had to cross horizontally through very difficult, tightly tree-studded, very steep slopes. It was thirty minutes of warm-up that left us heated to boiling temperature and soaked in sweat.

I also worry about my knees holding up, about wind-blown snow and death crust, and especially about skiing with thirty-year-old Austrians born on skis who have skied all year in preparation for this. I ask myself: "Okay, I'm here, but would it not be better to just rest among the trees, sense the majestic mountains that nature has made so incredibly beautiful, breathe in the fresh air, enjoy the sun, the fog and clouds, the liberating silence? Why give all this up for superhuman runs down mountain slopes sometimes so steep that I can touch the ground by just stretching out my arm sideways? Why forget that I am old enough now that at home I won't even jump down three stairs, and take on snow-covered rocks, tree stumps, and cliffs ten to twelve feet high, trusting that the deep powder will buffer the impact and loosen up my frightened, strained muscles, open up the pores to my heart and mind, drive oxygen down the long-unfed labyrinth, unite the brain with the body, and bring the sleepy awareness of a desk man back to the instinctive reaction of ancient animal ancestors. "Perhaps," I answer, "because this is one purpose of this truly remarkable activity. One can come close to and sometimes sense the great moments that poets and musicians have when struck by ideas so strong that they precipitated into works of art, leaving tracks for the rest of us to follow for generations."

Here in the snow you can feel either great peace or an odd displacement. Standing under the snow-covered trees, you can hear poems resounding in you, and, while skiing, actually sense yourself sliding through fables. At certain points you will feel goose bumps all over, which return later in summer when you think back on those moments. And, to balance the serenity and transcendence, you will hear, as you wait for the helicopter, many jokes, often terribly dirty ones, from the large repertoire of the forty people.

Sometimes my fellow skiers remind me of the behavior and psychology of the people with whom I was imprisoned, constantly carving out for themselves a little advantage, like the prisoners who always grabbed for the biggest piece of bread or tried to justify their actions in curious ways. Some always ski right behind the guide, some avoid carrying the backpack, some always take the best seat in the helicopter, some refuse to baby-sit skiers who fall. I will always remember the accident-prone orthopedic surgeon who claims, apologizing like Frau Hochwasser in the air raid shelter, that each of his many problems is a unique, unrelated occurrence. Another older man always wants to continue skiing even though the younger ones are worn out and ready to quit. Ironically, he is not one of the better skiers; in talking, he displays a childish mind while his dentures clap as if in danger of falling out.

Back at the lodge, after skiing, we share our pasts and presents. As soon as my companions find out I am originally from East Germany, they ask whether I have been back, what I think, and whether I want to help solve the problems left by the system's collapse. I tell them I was back in Jena right after the Berlin Wall came down, and have visited again many times. I have seen the towns and villages that looked then as dismal, dreary, and gray as if they dated from the Dark Ages. I describe the listless, almost prisoner-like faces of this population that was locked up so long. I tell how in Berlin I went back to look at the *Stalin-Allee*. In 1961, when Stalin's name was being removed from history, the avenue became *Frankfurter-Allee*. It is not what I thought it would be. After only forty years, its shabby apartments need repairs, and scars on the facades of the decaying buildings show where the facing stones have fallen off into the street. The showcase of socialism was shoddily built and poorly maintained.

When I went back, I contacted all the people still alive who had studied with me, but I could never go back permanently to help: basically, the people there don't want to be helped, especially not by a wise guy from the West. After the Wall came down, Zeiss, Jena shrank like a deflating balloon from a company with 32,000 employees to one with perhaps 800. Most East German industries suffered the same fate. East Germans are bitter about that because they felt they had given their best and were and are just as smart as people in the West.

What they don't fully realize are the long-term consequences of the socialist *Plan Wirtschaft* (planned economy), in which the government decided which company was going to produce what. No purchaser could choose among competing products: only one company was allowed to produce that product. Trade agreements and western embargoes forced Soviet-bloc nations to buy only from each other, always from the same companies, a de facto elimination of competition. Each company had a guaranteed market and could ignore any complaints from users of their product. There was no pressure to improve products, and the regime's compulsive secrecy about every aspect of technology and science meant that communication and access to information deteriorated, forcing even the best people to work in isolation. Cut off from the mosaic stones of updated knowledge that can be acquired in informal conversation or at international congresses and meetings, they had little chance to see the larger picture of any given technological puzzle as early as we did in the West. They always had to rely on secondary information, looking at western patents after they had been issued, or trying to make sense of bits of information from spies, not unlike piecing together scraps of paper from the garbage can. Finally, every year their companies had to send profits back to the central government, which redistributed them according to need—putting non-performing companies in the best position to receive the most capital. In addition, since the "right to work" was guaranteed in the east-bloc countries, everyone was assured of receiving his paycheck even if he had done nothing at work for months. Over fifty years, strong attitudes about work devolved into weak ones. No matter how much pride an individual had in his work, people tended to accept the status quo even after it had become obvious that what it was producing was not progress but steady deterioration.

While few of my friends ever joined the Communist party, they lived in that world so long that they eventually accepted this abnormal way of living as normal. When I was fighting as an entrepreneur in Silicon Valley, I often thought back to my friends left behind in the East, and sometimes I was envious. They could walk the trails around Jena I loved, and work in a protected job environment. Of course, they knew the government stole their earning power and planned it away or gave the goods they produced to the friendly Soviet Union, but everybody either got used to it or concluded they could do nothing about it. They were not pushed to build houses: there was no building material. They did not have to dress in fancy clothes: there was nothing to buy. They just had to wait their lives out.

During a reunion, I had the chance to observe my former classmates and was surprised to find among them the same associations as before, all affinities intact, each person's potential confirmed, personalities and traits and thinking patterns almost the same as they had been forty years ago. The ones who had stayed in the East had great pride; but isolated there, they never realized they were fighting with wooden bazookas in a fake economic environment. Lacking the competition of the capitalist world, they had been on teams that played ball without an opponent. They had lived in an environment where false reporting about output and performance was common and had few consequences. They had no idea that in the West, the slightest drop in performance could weaken one's ability to compete, leading to potential disaster and bankruptcy. They had lived in a surreal world.

Recently I sent a letter to a friend who had remained in Jena, asking him what he had thought at the time of my imprisonment. He was never a communist or a person with much liking for the East German system, but his answer gave me a lesson in just how twisted a sense of justice prevailed in the East and how the assumptions of the environment shaped the thoughts of the people there. He wrote me: "Even though you acted irresponsibly by trying to escape from the republic, and in light of the fact that you talked another engineer (my fiancée, Erika) into leaving, we thought your six-year sentence was too high." His response made me realize that he had lived most of his life under governments that saw no reason to relinquish part of the state's power by allowing citizens to have human rights. Living under such a government, even those individuals who did not support the ideology preached and practiced by the state found it almost impossible to reject completely the state's basic assumptions.

How, I asked myself, could one explain to these East Germans the difficulties faced by the famous *Treuhand* agency—the organization assigned after reunification to privatize East German industries—in trying to sell, merge, or clean up the dysfunctional industry of an entire country? This was an incredible undertaking, something that had never been attempted before. To perform such a sensitive operation without mistakes was impossible. In the capitalistic West, many companies had tried to turn around failing enterprises or just merge one company into another. Even on a scale so much smaller, many of these attempts had failed. How could the East German people understand all the mistakes and miscalculations, the acquisitions by West German firms for the purpose of eliminating competition, and the reductions of overhead and restructurings? How could they understand the resulting layoffs, except by feeling that they had been taken to the cleaners? How could they cope with the profit motive of the Western world and adopt the different economic assumptions of capitalism? How could they conceive new products matched to an emerging world economy and learn quickly the necessary marketing techniques? How could they know about the critical mass necessary to get new industries going, to sprout new companies and new work places, and to drive technology forward, as exemplified in Silicon Valley? Considering all this, to get the East Germans started was more difficult than getting all the sunning retirees in Florida to rise from their lawn chairs and park benches to build new industries—however old they were, they had at least lived their lives with the realities of capitalism. These are some of the reasons why the West Germans will have to pay for a long, long time without making the majority of the former East Germans happy. These were some of the thoughts I expressed to my ski buddies over dinner.

When they asked about West Germany, I mentioned that I was visiting Germany often and told them how I admired, but also wondered, how German enterprises managed to carve out their profits despite strong unions like the I.G. Metal demanding and obtaining the 35-hour week, six weeks vacation each year, twice as many holidays as we in America have, and total job and social security. In addition to all these benefits, the union gets raises across the board in addition for all its members, even though the members include a whole spectrum of crafts and skills, ranging from metal-working technicians to pilots. In fact, German enterprises had what seemed like a marriage contract with their employees: they could not hire and fire employees as we do in the United States, and the employees could not leave at will. Amazingly, almost every German could afford to travel to faraway places and spent a lot of time enjoying himself in beer gardens or even in oyster bars drinking champagne, with a warm feeling that a high and secure retirement income is waiting and that he or she is fully protected against financial hardship resulting from unemployment, illness, and disability.

All this gave me the impression West Germany has in many regards miraculously succeeded in establishing socialism, democracy, and freedom while the former East German regime had failed with all of three. Moreover, and curiously, after dissolving the socialistic East German regime, West Germany had integrated my former countrymen (all 15 million of them) into its system, providing them all with the same generous social benefits West Germans have enjoyed.

In general, though, biased as I was and am by living in the United States and even more by working in Silicon Valley, I felt that many Germans seemed unenthusiastic about life and not very cheerful in spite of all that good living. The once-thriving population seemed to be, all too often, waiting and expecting the government to keep things going as things had gone in the past, and to solve all problems, including high unemployment, without particular attention to what was happening in the rest of the world.

When I talked to West Germans about how people in the United States looked after themselves, most of them said they did not want to live that way. When I tried to tell them about Silicon Valley, which when I arrived was largely a valley with orchards, I found they generally believed most business ventures there had gone sour or had gone under. I heard many assert that our entrepreneurial approaches did not apply to Germany. While I thought that was true, it was very interesting to me that few wanted to change the system and consequently be exposed to higher risk. Most did not understand that in the long run they were taking even bigger risks by not changing anything, and that most of the failures in Silicon Valley were part of the natural process. They could not see that most failing ventures just made room for better ones that quickly absorbed any person laid off; that many very big companies, employing tens of thousands, had emerged, such as Hewlett-Packard, Intel, Oracle, Compaq, Genentech, Digital Equipment, Cisco Systems, and Applied Materials; that thousands of healthy little and medium-sized companies had made it; that many of those might fuel further progress after being absorbed into others; that hundreds of thousands of employees have become beneficiaries of change-induced prosperity because they own stock or have option packages to buy stock at a fixed price in the companies they work for; that freeways are packed with people wanting to give their best; that during 1997 alone 53,000 new job positions were created and filled; that a full 23% out of a population of 2.3 million were born in foreign countries, as I was; and that there is enough excitement in the air to keep people like me wanting to help, instead of thinking of retirement. When very major phenomena like Silicon Valley develop, and nothing like this is growing in your country, it should make you think. Perhaps European unification will free up economic forces and make talented people mobile, adventurous, and driven enough to make a Silicon Valley happen in Europe.

[1] Death crust is frozen crust up to three inches thick that sometimes forms on top of snow. A skier can easily break through at any time, and often does on each turn. Skiing on it is almost impossible—one tries to survive without falling badly. Skill and strength help, but everybody falls at times.

26

I am grateful to have met all my East German friends again. I still like these guys. Even though many of them are gray or bald, they seem unchanged, jewels I had left behind and have found again, representatives of my youth who continued to live in my German past. I was interested in what had become of them. Many had remained in East Germany; others had escaped like me and had spent their lives in West Germany. I was fascinated to learn how far each had come in life, how he viewed it, how satisfied he was, and what lessons he had learned. It was in many ways surreal to see them again because their lives gave me an idea of what I might have become if I had remained in East Germany, or if I had stayed in West Germany and not become an American. This was my own private study group.

My classmates fared similarly in many ways, regardless of whether they lived in East or West Germany. All of them worked all their lives to raise their children. Most of them still lived in two- or three-bedroom rented apartments, and had spent money carefully and lived within their means; few of them had saved or accumulated a lot. Looking at them, I got the impression that taking care of their families, fighting for existence, trying to fulfill their ambitions, and experiencing success and defeat had worn them down. Of course, everybody tries to appear still vital, but they are beyond the end of their careers, there is no more youthful force, and they have exhausted the energies that when we were young seemed limitless.

I was surprised at how well they had chosen spouses, and how in the course of a lifetime these women had become one with them. Their relationships seemed different from those I am used to in California, where the matrimonial glue is rarely strong enough to hold couples together.

Even though I had lived in a foreign country with a life so different from theirs, I felt still like one of them for the few days I was able to spend with my friends. It was as if little time had passed, not forty years. I needed them in the same way I needed the people heli-skiing—because, as I had realized some time ago, I had lost contact with most of the people I once knew. This was brought home to me one day when I was driving my open convertible down Highway 280; a wind took all the leaves of my dilapidated address book and blew them all over the landscape. At first I was shocked: I could not call my friends anymore. In time, though, I was even more shocked to discover that almost no one I had met, loved, dined with, or done business with over the last thirty-five years ever called me. Instead, calls came from people I did not know, like unsolicited junk mail.

After that I sorted through the more than 5,000 business cards I had collected and saved over the years. They represented the precious human resources I had encountered, people who had once been very important to me. Now I threw most of them away. Losing the rest of these acquaintances and friends seemed surprisingly natural. They were like out-of-date periodicals. Living as a naturalized foreigner in America, I had long ago realized that almost all relationships here were based on a purpose, and when this purpose was gone, they faded out to make room for new ones.

I shed no tears for those 5,000, but I cried when the Berlin Wall came down. This moment, in some sense, closed the books on the hard years I had spent in communist prisons, the pain I had suffered in restarting my life in the West, and the more than thirty years of separation from my relatives I had endured. Now I could at least go back.

I cried again when I re-visited Jena for the first time: I heard the people speaking my hometown dialect; I also saw how loveless and uncared-for everything looked.

Often memories and thoughts like this pass through my mind when I go for walks, or at night before I fall asleep. Consciously or unconsciously, I am searching for better solutions, or thinking about what I could have done better, what I have not done yet, and what I could do in the future. After fifteen years, I can look at my technical achievements at Optimetrix with detachment. Like the paving stones in an ancient Roman road, they are only blocks in a long and wide technology thoroughfare that have allowed others to advance farther or get ahead faster. I am now at an age where I can imagine my future as an older man, living in my own world, participating less and less actively in the day-to-day battles of the real world, just watching most of it go by like a movie, as I once did in the Astoria Theater when I was six years old.

But I am not there yet; I am here in British Columbia taking anti-inflammatory tablets and vitamin pills to help my muscles and joints recover from today's strain and to prepare for tomorrow's. At first I had thought I was the only one cheating nature that way, but I have now learned that almost everybody else is doing it, too. As we learn more about our bodies and find new cures, we use them massively to feel or be better.

Just before this trip, I read that a sheep has been successfully cloned in Scotland, and that scientists in Oregon expect to clone monkeys this very year. That makes my DNA project of 1976 look less bizarre. After only twenty years, we have already come halfway.

I think nobody will be able to prevent human cloning. First, specific genes will be changed; then organs will be cloned; and finally, someone will clone whole humans. Laws or not, there will always be a place where it can be done, a corner of the world where cloning will be allowed or will be done secretly in a closet. I tell myself not to worry too much about all this. By the time human clones arrive in our lives, many of us may have already altered our own genes to eliminate diseases or undesirable inherited defects. While too many clones might affect the variety of traits we get by heterosexual gene mixing, I think the human race will manage to control this. But clones must have the same rights as humans who are "normally" conceived and borne by mothers. God forbid that we should consider them something else or start a new discrimination and hatred based on genes, like the old ones based on race, religion, ethnicity, or national origin. Six thousand years ago, before the veil of refined religious ethics was put over most of mankind, cloning would perhaps have appeared more acceptable and less troublesome. People would have used the process to give new lives to the powerful, and the clones of many famous people of ancient times might still be living with us. The Egyptians would have cloned their kings and never built the pyramids, and Christians might have saved a sample of Jesus with which to resurrect him physically.

When I conceived of the DNA storage project, I thought it would be many years beyond my lifetime before humans could be cloned, and I thought I would never have to face the ultimate difficulty of this project—how to transfer conscience, soul, and, most importantly, ego into a person's clone; and even harder, figure out where to park the conscience, soul, and ego that would certainly develop in each clone as he or she grew up. In 1976 I never went that far in my thinking. I just wanted to help people, when leaving this earth, feel better by knowing that at least a small piece of their living matter would be preserved for a while. But as things look now, older people who have lived well and love life may strive to be transferred into younger clones while they are still living, thus continuing their life as younger persons. That, of course, brings the ignored, unsolvable aspects of my 1976 project right into the present and makes me shy away.

One thing is sure: nobody can assess the full impact that biological technology will ultimately have on the human race. When Bardeen, Brattain, and Shockley invented the transistor in 1947, it was originally thought of only as a replacement for the vacuum tube that we all know from older radios and TVs. Even though everybody knew then that the transistor was easier to manufacture, much faster, and at least one million times smaller, nobody guessed it would be the start of a new information age—nor did they foresee its consequences.

But I have to put all this aside for a while; it is morning, and the wake-up bell is ringing to announce the stretch class. Opening the curtains, I see the almost overwhelming winter landscape with glistening mountain peaks and forests deep in snow along the long valley below the lodge. The fresh snow has no tracks. No road leads here; this lodge is only accessible by helicopter.

Outside, it is very cold—minus 10 Fahrenheit—but the weather is beautiful. That gets everybody in a good mood. The skiing will be fantastic. After a very enjoyable breakfast, the head guide gives us the weather report and announces which mountains we are going to ski. It is time to get boots, down suits, heavy gloves, sunglasses, hats, and the broad powder skis that will keep us from sinking too deep into the snow.

We are ready. I hear the helicopter warming up its engine. Soon we will feel as if we are in a war game, with all of us huddling down on the copter's pick-up side to avoid its deadly rotating blades. We climb in, sit as tight as herrings in a box, and listen to the noise as the powerful motors lift us off the ground. The helicopter turns in circles and shoots down the valley, flying to battle with the speed I saw planes use in the dogfights in World War II, and then climbs toward the first drop-off spot. We cannot talk much because of the noise, and can only give each other looks or thumbs-up signs. Although my fate is in the hands of the pilot and the functioning of this machine, I feel safe in the helicopter. I am satisfied with whatever comes; having already lived a full life, I am here just asking for a bit more gravy.

When we are close to the top, a sharp wind from the other side of the mountain drives the helicopter up and backward, almost toppling it over. The pilot handles this by pushing his stick hard forward, and we go down on the landing site where the heli's two skids dig deep into the snow, just six inches off the little red marker flag.

We climb out, and the guide unloads all our skis from the basket on the other side of the helicopter. We give the pilot the ready sign, and he takes off. We stand and watch the machine disappear and soon cannot hear it anymore. The world up here is silent once again.

Another guy and I go to work to dig the skis out of the snow, calling the name of each skier and handing them out. It does not take long, and we are all ready to ski. Before we can start, we have to pick our partners for the day. I am not fast enough and end up with a young Austrian who seems to want, like me, just to ski alone. He is in his early thirties. With his full, long beard and long hair tied into a ponytail, he looks like a violin-maker from the Oberammergau, ready to play a part in a passion play. I have watched him before and know that he skis incredibly well.

He says "Hello" as if we had not met before. I see a touch of worry in his face—I know he is afraid I will slow him down. I feel some concern in my gut, but I will give it my best; I don't want to be left out.

Then we start out, one after another, in a long queue behind the guide. The snow is very light and deep; this promises to be a very good run. The guide stops frequently, studying the snow and collecting the group. We cannot afford to lose anybody up here. Even if someone only falls into a drift, he might not be found. It looks as if it is starting to snow, but perhaps the wind is only blowing snow from the other side of the mountain.

Finally the guide says, "From here on, you can pick your own pass to the valley. I'll see you down there. Don't go too far to the right—we are still on the glacier. Farther down on the left there is a cliff. Stay with your partner and use the leapfrog technique so nobody gets lost. And, most important of all, don't ski the very steep sections or you may cause an avalanche."

Before I can think, my partner takes off straight down where there seems to be only air. I have no choice and follow him. He is already forty yards ahead of me, hardly visible in the snow he stirs up. I go straight; then, once I am closer, I figure-eight him by crossing tracks in the middle and above each turn that he has made. That way I don't have to think where I am going, which helps in timing my turns, and I can go faster. Soon I am right behind him and know that he must be surprised that I am there. He goes to the right, to the best feather-light snow. Over there I can see the bluish-green walls of the glacier. They remind me of the crack I was in, but there is no time to think about the dangers that lurk there, and I follow my partner, matching his rhythm and speed.

He should have stopped five times by now, but he keeps on going, not caring that I can hardly see through my ski mask because of all the snow that has collected on its glass. I keep copying his smooth rhythm, can ski like him now, feel no pain, just incredible excitement and pleasure. Then in a long, steep, tree-bordered slope of what must be a meadow, I claim the lead due me by going straight down, passing him. What jubilation in me! What a sweet moment! Oh, how I wish my friends could see me! How can I describe this? I am in a rage and feel like a winner. Sparks of youth are falling right into my soul. Now I can go back into turns while he stays right behind me. I am showing this guy what I can do, that I am still there, that you cannot count old guys out too quickly. I am skiing in front of my ideal, without falling.

Suddenly I see whole sections of snow to my side moving much faster than I. Avalanche! The realization strikes me like lightning and sends shock deep through my bones. We have had it! This is the end! In split seconds we will be in twisting, fast-moving snow—seeing only white as we fall, float, and tumble in a furious hell until we get smashed into the trees or squeezed into tightly-packed bundles, deep under the snow with lungs gasping for air and only seconds to live.

There is no time for regret, no time to change our fate, no way to escape. We have begun the transition to death and will soon see the other side of this world.

But we are only half out of this world; determined to escape for now our ultimate fate, we are lucky enough to ski out of this madness of pulverized ice up to the safe sloping side of a gully, from where we two escapees from death watch in terror as the thunder-mountains of snow blast their way through the trees down into the valley.

Exhausted and shaken, we let ourselves fall into the snow to rest and to recover from our scare. "Maybe you should not ski anymore," I say to myself. "This was dangerously close! And what would your family do without you?" After a while, when everything appears to be calm and silent again, we ski farther down to where our companions have been anxiously waiting for us.

After six or seven runs it is lunchtime. We always eat it outside. The guides know all the beautiful spots, and whether we have sun or snow with it, lunch is fantastic. We eat standing up or sitting on our skis, and talk a lot. Even here in the snowy wilderness there is always somebody who cracks jokes. This time it is a bad one about O.J. Everybody thinks he murdered his wife, but the verdict left a doubt—just as, when Hilda was raped in East Germany, the authorities insisted on describing the rapists as "persons in Russian uniform" rather than "Russian soldiers," creating an official doubt (though none existed in our opinions) that left everything up in the air, with no one guilty.

Later in the day, while we are walking in deep snow over a long, flat glacier, we look over an ocean of suncrust created by very cold nights followed by sunshine. The whole surface seems to be tightly seeded with diamonds as billions and billions of glittering ice crystals scatter and reflect the sun's light. I am almost overcome by its intense yet peaceful beauty. What a contrast to the deadly ride this morning, when I came as close to the end as I had on that day with Erika so many years ago when I rode the avalanche in Austria.

I had visited Erika last year and showed her some photos of my family. She had looked at my younger son's picture peculiarly, as if she wanted him. She had only daughters. At that moment I wondered whether she had ever found again a love like ours. I had always longed to relive the time with her, that time now so far in the past. Erika had looked up then and asked, "Wouldn't it have been better if we had stayed together and not left Jena?"

I had often pondered that question, and could only answer, "I don't think so."

Had we stayed, would our great love have continued forever? There are no answers for such questions, and one cannot change or return to the paths one walked.

Each time I think about my past, I long to see my former hometown again. Even though I have become an American, I still have Jena's nature and make-up in my blood. One remains connected to one's past, and, inside one, everything from the childhood village or neighborhood—its feel, its smells, its map, its size—survives.

Looking back, I see that, in many respects, I am still a

homo Jenensis—the Jena man I once was. Even though I live now in a beautiful house and drive a Mercedes, I still like to eat only two slices of bread with some jam for breakfast, as I did in my mother's apartment so many years ago. My feet are still on ground that is, to a great degree, Jena's. On Christmas Eve I wish I could feel again as I did when my mother called us to receive our presents and I saw for the first time the little Christmas tree in the glow of real candles. On New Year's Day I would still like to yell out cheerfully my *"Prosit Neu Jahr!"* ("Happy New Year!") to neighbors across the street, as I used to in Jena. But if I did, my neighbors might shake their heads and think me crazy.

My Jena soul has been modified, however, by countless adaptations to a new, far more complicated life in America. New attitudes have taken root in me. Like so many people these days, I am sometimes left a bit disoriented by sensationalistic press accounts, agent-fabricated entertainment news, and seductive advertisements, all of which leave the truth somewhat fuzzy. As when the Russians changed the past, or when an accountant cooks the books, or when a director makes a powerful film that imprints its false version of history into our minds, after a while nobody knows anymore what the real events or the correct numbers are, what was altered and what was not, or what real justice is. I am afraid that factual fuzziness and media exaggeration have already affected us all, distorting our perceptions and judgment, leaving us unable to interpret or react clear-headedly to real stimuli. We have become acquiescent, indifferent, callous—our souls changed forever.

I am apprehensive that so much being broadcast is about murder and crime; that so many people spend so much of their time watching television, switching channels to wring out more information content—and perhaps from their lives; that for so many the stock market reports have become the most important, real, and truthful programs; that the world is being reduced to an economic abstraction in which globalization demands everybody's full attention, sets new standards, and establishes new ways of life; where ancient diversities and cultures get lost and are replaced by a uniform new culture; where we float upon oceans of information into the future and have to face challenges without proven leaders or proven ideas for dealing with them—somewhat as the West Germans had to face remaking a whole country.

We don't know what the world will look like even a hundred years from now. But we are driving full speed into the future even though we have no road map and can see only the short stretch of road just before us, not knowing whether a bridge may be missing or a truck is stalled right after the next turn.

What will life be like after individual countries have become unimportant and each person is equal, when the information age has fulfilled its cold promise and everything and everybody is connected and administered without people, when all records, obligations, and rewards are guarded and dispensed by computer, after whole armies of administrators and politicians have become obsolete, as whole armies became obsolete once the Cold War ended?

Who knows what may happen to our genes if we continue developing our technology? We don't yet understand how, when, and why mutations occur in the human race. Can the sheer mass and quality of what we are doing in technology trigger a quantum leap in such aberrations, or will we become smart enough in the future to initiate and execute positive genetic changes to match ourselves better to the devices we may create?

But enough of this: I am here to ski these beautiful virgin slopes. From the helicopter I can see the curly tracks we just made, leading all the way from the top of the mountain down the tremendous steep slope to the valley. The sight makes me feel great. I am still ready and still have enough power to face the world. There is so much left to be done.

We ski until late in the afternoon. And as we are flying one last time toward the snow-laden eerie mountain tops, my thoughts return to how close I came again to the end, how even this ride might be my last, and how at any moment my precious spark of life could leave me.

Sitting so tightly packed in the helicopter, I am suddenly reminded of my trips around the world. Back then, when I left my comfortable life at home and was traveling to far-away places with many anonymous people on tightly-packed planes, buses and trains, I felt strangely like one of them. For safety, I wore very little, traveled without a suitcase or even a watch, and could imagine myself on my final journey.

Dying might just be that way. One travels without friends and belongings, one has no reason to pretend, no need for masquerade. One just travels onward, or flies like us over the glaciers, and all of us are finally the same. There is no more thinking, "Is this all there is?" No more raking the ground where painful weeds lie buried, no more tears when thoughts of suffering reappear. No more striving for lust and fulfillment. No more asking for the meaning of life. No more questions: How much is enough? Am I closer to zero or infinity? What is the truth? Who made the universe, and where does it end? On one's way all that remains are good memories glistening like the diamond snowflakes on the glaciers below.

ACKNOWLEDGMENT

I am greatly indebted to my wife, Irene, who quickly and tirelessly typed all handwritten parts of my manuscript into the computer while encouraging me constantly with remarks such as: "I want more" and "this is going to be a movie."

I also express my gratitude to my Editors Dr. David Brownell, Peter Jones and Larry deStasi who gave me much support and encouragment throughout the writing of the manuscript and did a marvelous job in editing it.